F IBER OPTICS
AND
CATV BUSINESS STRATEGY

For a complete list of the *Artech House Telecommunications Library,*
turn to the back of this book . . .

FIBER OPTICS
AND
CATV BUSINESS STRATEGY

Robert K. Yates
Nolwen Mahé
Jerôme Masson

DTI Telecom, Inc.

Library of Congress Cataloging-in-Publication Data

Yates, Robert K.
   Fiber optics and CATV business strategy / Robert K. Yates, Nolwen
Mahé, Jerôme Masson.
      p.   cm.
   Includes bibliographical references .
   ISBN 0-89006-413-X
   1. Cable television. 2. Fiber optics. I. Mahé, Nolwen.
II. Masson, Jerôme.   III. Title.
TK6675.Y37 1990   90-30132
621.388'57--dc20   CIP

International Standard Book Number: 0-89006-413-X
Library of Congress Catalog Card Number: 90-30132

10 9 8 7 6 5 4 3 2 1

Table of Contents

Preface

Cable television companies can build high quality and reliable systems using fiber optic technology. A strategic backbone infrastructure for new broadband services can be put in place for an incremental investment. This infrastructure can be built with an investment of approximately $100 per subscriber.

Fiber optic deployment in cable television is generally triggered by the need to provide an increased number of channels; usually going from 36 to 60 channels. To optimize deployment plans, a system operator should set strategic objectives and use these to drive the selection of new technologies and system architectures. A typical set of strategic objectives could be:

- improve picture quality to support high definition television,
- improve system reliability,
- build a physical plant compatible with future services (e.g. bidirectional data transmission).

Architecture and technology choices should be optimized using these objectives as constraints in the planning process; i.e. an optimal configuration should be found which meets these objectives. This may not produce a minimal cost solution, but it will produce a "local" minimum in the context of what the system operator is trying to achieve.

Architecturally, cable television systems should use the strategic investment in fiber optics to minimize the amount of coaxial tree and branch plant. While cost effective and efficient technology for large numbers of broadcast services, coaxial plant is not optimal for narrowcasting, pay per view, or bidirectional data or video services. Cable system operators should "push" fiber optics as far into the distribution plant as is affordable using their system's strategic objectives as the driver.

Technologically, cable systems can greatly benefit from deploying fiber systems based on amplitude modulation techniques. AM systems can prove in against a microwave radio upgrade when a channel upgrade is being considered. In addition, AM fiber systems provide a low cost alternative for distribution systems. While today's specifications on AM fiber are perhaps not all the industry would like, in many cases performances will be better than that of coaxial cable. Future developments which support AM techniques are likely to stimulate further improvements in the specifications. While not having the ruggedness or long-haul capability of FM or digital systems, for relatively short distances AM fiber optics offers an inexpensive approach to building cable systems.

The future of fiber optics is wide open. The "infinite bandwidth" of the fiber is just starting to be tapped. Strategic areas of research and development are already yielding interesting systems using, for example, external modulation or new optical splitting techniques. Systems using these approaches will start to become available in the next one to three years.

With fiber optics there is an opportunity to simplify the engineering of cable television systems. This is accomplished by setting technical goals for the system and then varying technology choices to find ones which meet the goals. This "budgeting" technique for technical parameters assists in making the fundamental choices. Ensuring that the initial objectives are well understood will, in turn, ensure that deployment plans remain consistent as the business and technology evolve.

Cable system operators should start reaping the benefits of fiber optics while building in capabilities to support new technologies and services. Immediate benefits to the cable system of fiber optics are reduction of amplifier cascades to improve reliability, and enhanced system performance and monitoring.

For future service capabilities, cable operators should ensure that fiber cables are sized appropriately in order to accommodate new electronics on the ends within the planning horizon.

Acknowledgements

This book is a revised version of DTI Telecom, Inc.'s "Fiber Optics in Cable Television - Planning Guide", published in May, 1989.

The authors wish to acknowledge the generous support of the Canadian Cable Television Association and in particular Roger Poirier, Vice-President of Planning and Technology. In addition Robert Ménard, President of DTI Telecom Incorporated, provided invaluable guidance throughout the project.

Throughout the text we have tried as much as possible to cite references and indicate sources of the material. If we have missed anyone, we apologize. We are very grateful to the numerous vendors who were willing to provide data or enter into detailed discussions on some of the finer points of cable television systems.

R.K. Yates
N. Mahé
J. Masson

Montreal, Quebec
January 10, 1989.

The Authors

Rob Yates holds a Master's Degree in Engineering from the University of Toronto and has over 10 years experience in various fields of telecommunications. While working for a major equipment manufacturer, Mr. Yates was instrumental in formulating applications strategy and product definition related to fiber to the home, broadband switching, Synchronous Optical Networks (SONET), network survivability and operational support systems. Mr. Yates is currently Director of Studies at DTI Telecom, Inc., a Montreal-based consulting and engineering services company.

Nolwen Mahé holds an Electrical Engineering Degree from INSA, France, and a Specialization Certificate in Telematics, Telecomputing and Networks from ENST, Rennes, France. She has been involved in research and development projects concerning network planning, simulation and routing. Ms. Mahé is currently a consultant in fiber optic planning at DTI Telecom, Inc.

Jérôme Masson is a Professional Engineer in Electrical Sciences from the University of Sherbrooke. After six years in the Quebec Ministry of Communications, he joined a major cable television company as director of engineering, following which he was director of operations for an electronics manufacturer. Mr. Masson is currently director for cable distribution at DTI Telecom, Inc.

Chapter 1
Why Fiber Optics ?

1.1 Cable Television Business Strategy

This chapter is not about fiber optics technology, although the words "fiber optics" do creep in. The CATV facility is a television "pipeline". The pipeline defines the business. Fiber optics is a new pipeline. The chapter is about cable television business strategy. The rest of the book is about the pipelines.

"Cable companies, traditionally a monopoly in North America, are going to face competition during the 1990s"[1]. This competition comes both from other telecommunications industry players as well as other technological competitors. A cable company caught between consumers buying advanced television systems and telephone companies providing digital broadband services does not end up with much of a business. There is an emerging competitive infrastructure which will squeeze the cable industry.

These industry forces are causing the cable companies to rethink their business. Cable companies are devising strategies based on:

• customer satisfaction and loyalty,
• growth beyond traditional services,
• potential telephone company competition [2].

The cornerstone of a strategy for business growth should be: customer satisfaction and loyalty. When asked what is most important to them, some 25-30% of subscribers would say "signal quality and reliability" [3]. This result has been developed by Cox Cable into a "hierarchy of needs" from the cable customer's perspective. "The highest priority and the foundation of customer satisfaction lies in the delivery of consistent high quality pictures and sound free from interruptions" [4].

Building the cablesystems of the 1990s will rely on a sound base; an infrastructure which meets subscribers' basic needs and which provides for a future business as advanced television comes along and regulations evolve to provide for more competition in the provision of "broadband services".

1.2 How to Use This Book

In addition to providing technical information, the bulk of which you will find in Chapters 3,4, and 5, the book provides a methodology for the evaluation of new technology. In each chapter, this methodology is illustrated using a model CATV company called "Hometown Cablesystem".

The methodology illustrated takes the business objectives for the cable television system, translates them into numbers and uses the numbers to develop a strategic plan. From there the engineering task will benefit in two ways:

• reduction in the options open for evaluation,
• reduction in the overall time and cost to engineer a rebuilding or channel upgrade.

Strategic Planning

Strategic planning replaces neither high level objective setting nor detailed engineering. This book should be used in conjunction with those activities to help optimize deployment plans.

The planning methodology itself, illustrated in the Hometown examples, is an iterative process of looking at technologies and system architectures. Using this approach, an optimal system configuration is found.
The six chapters of the book cover:

• Business strategies and objective setting - Chapters 1 and 2
• System architectures - Chapter 3
• Fiber optic system engineering - Chapters 4 and 5
• Operational considerations - Chapter 6

The user should take his system, set objectives for it, then iterate between architectures (Chapter 3) and technologies (Chapters 4,5) to determine the optimal plan. In the book we show one iteration of the technique for Hometown Cablesystem. To optimize the plan several iterations should be done. Once the strategic plan has been

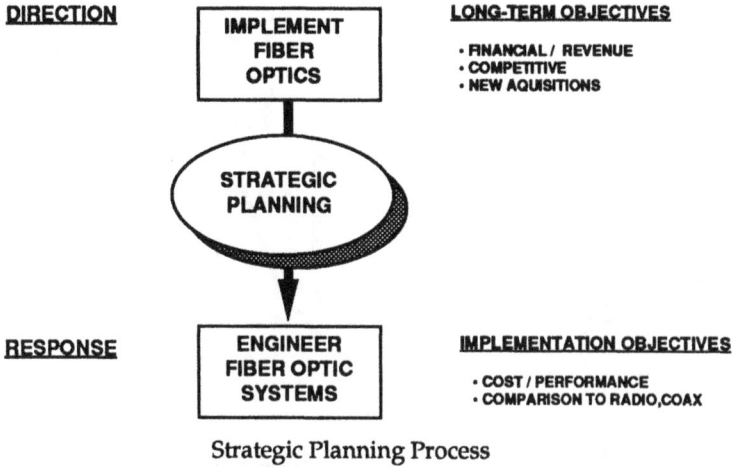

Strategic Planning Process

developed, the information in Chapters 4 and 5 also supports detailed engineering of fiber optic systems.

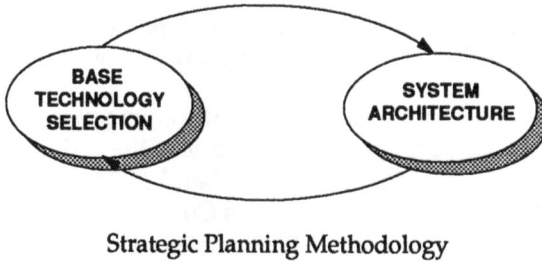

Strategic Planning Methodology

1.3 Channel Capacity

Channel requirements have continued to grow since the original cablesystems were pioneered in the early 1950s. With the recent addition of speciality channels and the increasing number of pay channels there is no indication that this trend will slow down in the near future. Typically North American cable systems are at 36 channels and are either in the process of upgrading to 60 channels, or will be in the next several years. This need to upgrade the number of channels is the predominant trigger for looking at new technology.

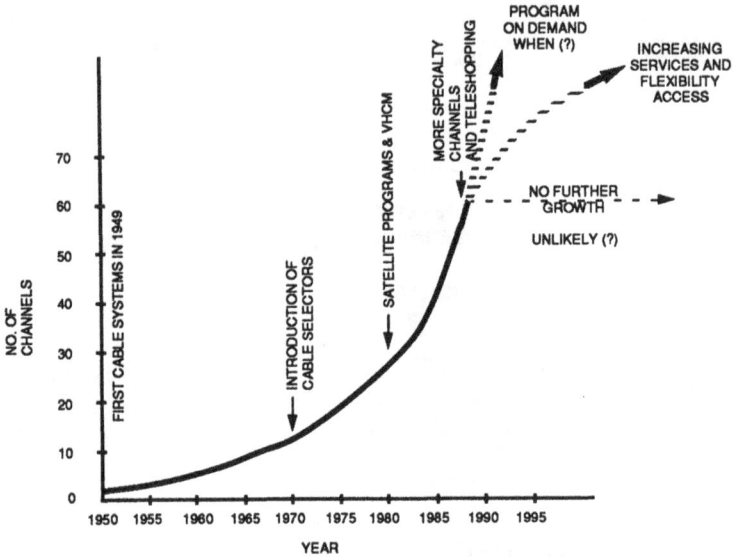

Approximate Growth in Number of Channels

(Source: "Study on the Application of Fiber Optic Technology to
Cable Television Systems", private study, DTI Telecom, Inc., 1988)

A second consideration in channel capacity is the emergence of advanced television. This will increase the number of channels due to the added variety which will become available, and also due to the increased bandwidth needed to transport the signals. For example, North American Philips Corporation has proposed a High Definition Television (HDTV) scheme which creates an HDTV channel by combining an enhanced traditional channel plus an "augmentation" channel [5]. Other schemes use varying amounts of bandwidth but usually occupy more than one channel. An important strategic issue for cable companies is that they avoid creating a bandwidth "bottleneck" into the subscriber. (Fortunately, fiber optics makes incremental bandwidth less expensive than traditional transmission technologies.)

A third consideration is the "trend to digitization" [6]. This is something already seen in digital watches, digital calculators, digital stereo, and digital compact disks. Digital coding creates a very clean, robust signal but requires a relatively large amount of bandwidth. The 6 MHz TV channel we know today in the analog world would become at least a 45 MHz channel in digital; that is, it occupies the equivalent of eight or more existing channels. Strategic plans will assume continued growth in channel requirements.

1.4 Signal Quality

Consumers are increasingly aware of signal quality. The consumer electronics industry has continued to improve the quality of the picture presented on television sets and VCRs. The more processing a TV set does to enhance incoming signals, the more the incoming noise becomes apparent on the screen. With HDTV, it will be even more imperative for the cable delivery system to be "clean"; high definition noise will not be much of a moneymaker.

An objective should be set for the cable system for picture quality to be compatible with HDTV transmission, as well as to provide the subscriber with a signal comparable to what he can get from today's VCRs and sophisticated TV sets. It has been estimated that 65% of CATV households have VCRs [7]. There is also a high coincidence of VCR owners and PayTV subscribers. Subscribers will continue to be "seduced" by the consumer electronics industry.

1.5 System Reliability

The other part of a quality signal is its reliability. Spending money on reliability is like buying an insurance policy. Each link in the system carries a certain number of subscribers, each of which account for a certain amount of monthly revenue. If you consider that some of that revenue may be lost due to outages then you can get a gauge on what you should pay to make the system more reliable.

Example:

If 170,000 subscribers pay $15 per month for service, there is yearly revenue of $30,600,000. If a typical insurance rate of $1 per $1000 of coverage per year is used, I should be willing to pay $30,000 per year to insure the revenue. Capitalized over five years, this would equate to a first cost of about $11 per subscriber. Based on this one should be willing to pay at least $11 per subscriber *just* to make the system more reliable.

1.6 Long-term Strategies

While the cornerstones of the business are signal quality and reliability, near term plans for the cable system must not exclude longer term considerations. In the process of upgrading or rebuilding the system, the opportunity exists to add structure or capability to ensure that in the future new businesses can be supported. This may be as simple as adding in extra fibers to a cable when a supertrunk is being upgraded.

From a longer term perspective, there is a significant opportunity to build a cable system to provide advanced television and broadband services. Cablesystems can implement fiber optic trunking while doing a channel upgrade for $50-$100 per subscriber. While not providing "fiber to the home", the $100 investment level will put in an infrastructure to support some new services. When telephone companies look at providing the same services, they are generally looking at per home costs of at least $1000 [8]. This added cost comes from two considerations:

• the bandwidth of the drop cable in the telephone plant is designed to support telephones, not video, and therefore the cable has to be replaced;
• the systems are generally built to provide bidirectionality and centralized switching.

The objective for a cable system is to be incrementally aggressive and build a strategic infrastructure at around the $100 per subscriber level.

Telco Competition

Telephone companies have been conducting studies with results which show that in the mid 1990s (some claim 1992), the cost of laying fiber for new construction will be the same as that of copper, considering only POTS (Plain Old Telephone Service). Among other things, this induces the Telcos to investigate more thoroughly the capabilities of fiber optics in their networks.

The telephone company does not want to be in the CATV business. It wants to be in the broadband services business where providing cable service transport is one component. The other components are getting better defined in the number of fiber to the home trials, which are underway in the US. Southern Bell has systems from

AT&T and Northern Telecom. Bell of Pennsylvania is working with Alcatel, and GTE (Cerritos) is working with AT&T, GTE Labs, and American Lightwave Systems [9].

Despite these practical assessments, a number of open questions prevents any agreement on the appropriateness for Telcos of bringing broadband fiber to the home. Among these, we can quote the uncertainties of the network architecture to be used (in Chapter 3, you will find more details about some of the current approaches), the evolution of fiber and optoelectronics and important regulatory questions: whether the cable TV market is to be open to the Telcos, what would justify the cost of fiber installation and, in that case, who will pay for the upgrade, i.e. how will the revenue base evolve [10, 11].

How this will all come about from the technology and regulatory viewpoints remains to be seen, but the Telco is not something a CATV company can ignore. Whether to integrate with the Telco ("a connection of the cable and telephone company networks would build on the relative strengths of both industries, with the residential subscriber benefitting", is one opinion [12]), compete with the Telco or sell out to the Telco, should be a component of the long-term strategy of a CATV system operator.

1.7 Hometown Cablesystem

To illustrate the planning methodology and the resultant technology and architecture choices we have developed a model cable company known as Hometown Cablesystem. We have considered a part of Hometown which contains 70,000 subscribers served from a head-end 20 km away.

Hometown is a 600 km system, giving us about 117 subscribers per km. This is fairly dense, but Hometown is built on a point of land and has developed from the coast inland over the last 15 years or so. Our densest area is a radius of 5 km around the hub. The furthest subscriber served from the hub is about 10 km.

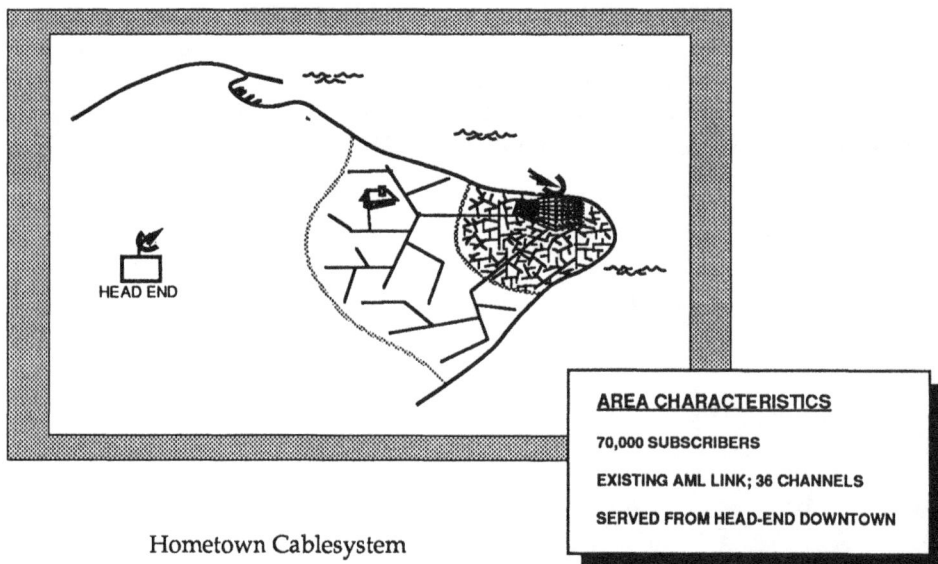

HEAD END

AREA CHARACTERISTICS

70,000 SUBSCRIBERS

EXISTING AML LINK; 36 CHANNELS

SERVED FROM HEAD-END DOWNTOWN

Hometown Cablesystem

Hometown Cablesystem was contrived to illustrate the approach to fiber optics and is not entirely typical of any cable system. It provides some numbers which turn out to be fairly representative.

In Hometown, the trigger for looking at fiber optics is a channel upgrade requirement. The system is currently at 36 channels and we would like to go to 60. The system has an existing microwave link from the head-end to a hub which is roughly at the center of the population density of the area.

Analysis of the Hometown system indicated that the channel upgrade would cost $70 per subscriber without considering any new technology. In addition, we have set the following objectives for the system:

• to provide a signal quality to each subscriber competitive with that of the Super VHS systems, and which would support the introduction of HDTV.

• to make our signal available at about the same level as the subscriber gets from his telephone i.e. about 15 minutes per year of outage time.

• to build an infrastructure for new services in the future.

It was decided not to go too far with new services in Hometown. There is a possible market for bidirectional data services but we do not know if the customers will want to lease those from the cable company. In our near-term plans we allocate a budget for "futures" and spend it appropriately to insure that these types of service are not locked out.

Having thought these objectives through, we are willing to pay an incremental $30 per subscriber to make sure they are met. This in addition to the minimum of $70 we already identified for the channel upgrade gives us a $100 per subscriber maximum budget. The plan is to implement an architecture and technology solution which will meet the objectives within the $100 budget.

References

1. Hamilton-Piercy, N. (Rogers Cablesystems), The Industry Imperative , *Cablecaster Magazine*, March 1989, p. 22.

2. *Project 94: Cable's New Horizons*, Canadian Cable Television Association, February 1989, p. 66. (Note: while these objectives are quoted from a Canadian source, the authors believe this is generally reflective of US and Canadian issues.)

3. Most cable companies have internal sources of customer survey data. These generalized findings are based on a number of discussions with Canadian and US system operators.

4. Cox Cable, San Diego; quoted in Hamilton-Piercy, op. cit.

5. Toth, A. (Philips Laboratories), High Definition Television - A North American Perspective, *Fiber Optic Communication and Local Area Networks Conference*, 1988 Proceedings, pp.331-337.

6. Ciciora, W. (American Television and Communications), The Trend to Digitization, *National Cable Television Association (NCTA) convention*, 1988 Technical Papers, p. 184.

7. Horowitz, E. (HBO), in National Cable Television Association (NCTA) Western Showcase, Anaheim, December 1988.

8. Hightower, N. (BellSouth Corp.), Economic FO System for New Residential Services, *Telephony*, March 17, 1986, p. 44.

9. There have been a number of published summaries of the fiber to the home trials. One good one is presented in Lightwave magazine, February 1989. Another reference is Kaiser, P., Fiber-to-the-Home, *Optics News*, October 1989.

10. Egan, B., Capital Budgeting for Fiber, *Telecommunications*, May 1989.

11. Scully, S., Telcos' Plans for Fiber Gaining at State Level, *Lightwave Journal*, August 1989.

12. Egan, B., *Capital Budgeting Alternatives for Residential Broadband Networks*, Columbia University's Center for Telecommunications and Information Studies (CTIS), draft paper, quoted in *Fiber Optics News*, November 13., 1989.

Chapter 2
Objectives and Economics

2.1 Fiber Optics versus Other Technologies

Chapter 1 is used to define top-down strategy to be addressed by fiber optic systems. This strategy can then be translated into numbers (dB,$,etc.) which will enable the system operator to make decisions on architecture and technology selection. This assures that near-term fiber optic deployment is compatible with long-term objectives.

There are two important things to understand about fiber optic technology. The first is that the world of fiber optics is somewhat "theological". Fiber optics is a fundamentally different transmission medium and combines the best of all worlds; unlimited bandwidth with minimum loss and no EMI or RFI issues. One of the industry's best known "gurus", Dr.John Midwinter of University College London, describes the potential of fiber optics as "awesome". He states that a single-mode fiber has bandwidth of 20,000 GHz around the 1550 nm wavelength [1]. According to our calculator, 20,000 GHz computes to around 3 million television channels. This figure is a part of the "theology" of fiber optics. It is important to understand the theology in order to believe in the long-term potential of the technology. Combining the capacity of the fiber with its loss characteristics (typically less than 0.5 dB/km), the traditional transmission equations for bandwidth and distance change; i.e. bandwidth will be "cheap", an important consideration for future services such as HDTV or video phones. Fiber optics can potentially replace most other transmission media, including coaxial cable, copper pair, radio, and, in some applications, satellite.

The other side of the coin, the second thing that one needs to understand about fiber optics, is its "mythology". This is illustrated by hyperbole in the popular press on the alleged current abilities of fiber optics. For example: "with a fiber optics pipeline, telephone companies can deliver many new services, such as home banking and pay-for-view television" [2]. The present cost of fiber optics to the home would triple or quadruple the investment in the telephone plant, and this is already significantly greater than that of the cable television plant. It is important to understand the long-term potential as well as the short-term limitations of fiber optics.

The final outcome has not been determined concerning what the ultimate fiber architecture for cable television will be, or how much a laser will eventually cost. While fiber to the home is not affordable today, by being "incrementally aggressive", CATV systems can make an investment in the new technology and build a strategic infrastructure for broadband services and communications.

Fiber Optics Is Not Just Another Technology Option

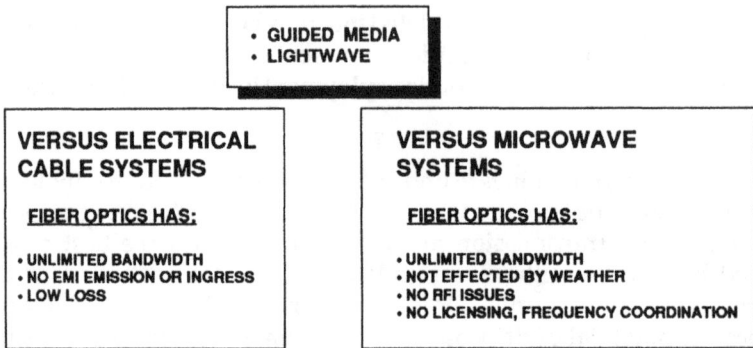

```
                    ┌──────────────────────┐
                    │ • GUIDED  MEDIA      │
                    │ • LIGHTWAVE          │
                    └──────────────────────┘

┌─────────────────────────────┐   ┌─────────────────────────────┐
│ VERSUS ELECTRICAL           │   │ VERSUS MICROWAVE            │
│ CABLE SYSTEMS               │   │ SYSTEMS                     │
│                             │   │                             │
│ FIBER OPTICS HAS:           │   │ FIBER OPTICS HAS:           │
│                             │   │                             │
│ • UNLIMITED BANDWIDTH       │   │ • UNLIMITED BANDWIDTH       │
│ • NO EMI EMISSION OR INGRESS│   │ • NOT EFFECTED BY WEATHER   │
│ • LOW LOSS                  │   │ • NO RFI ISSUES             │
│                             │   │ • NO LICENSING, FREQUENCY COORDINATION │
└─────────────────────────────┘   └─────────────────────────────┘
```

Fiber Optic Technology

Fiber optics can more accurately be described as "lightwave", or guided wave optics. Fiber optics is a guided media (i.e. cable-based) transmission technology with the properties of radio waves. Fiber optics is a young technology. Its "unlimited bandwidth" is just starting to be tapped.

2.2 Linking Long-term Goals to Short-term Strategies in Hometown Cablesystem

In this chapter, rather than trying to explain a rigorous methodology for objective-setting and economic benchmarks, the Hometown Cablesystem will serve to illustrate the approach.

In order to define a strategy for Hometown, we conducted a survey of system operators in the US and Canada.We were able to devise a set of deployment parameters which would drive the decision making process for fiber optics. The process of the survey and the results of it will illustrate the methodology of defining a strategy for the cable system.

Hometown Cablesystem Strategy - Survey

The survey was undertaken in four steps. Each interviewee was first presented with the Hometown strategy outlined in Chapter 1. The second step was to identify the amount of money the system operator would be willing to pay to meet the strategy. Thirdly, this amount of money was allocated over the various elements of the strategy and, finally, the technical parameters for the strategy were defined.

We developed a complete budget for Hometown which included the money we would spend and what we would spend it on. In this way we linked an overall business strategy to deployment plans. (Note: $ figures are generally assumed to be Canadian unless otherwise stated.) A summary of the survey and its results follows:

Step 1 - The Hometown Strategy

As defined in Chapter 1, the proposed strategy was:

• to provide signal quality competitive with the VCR marketplace;
• to provide system reliability competitive with Telcos;
• to position the business for bidirectional or other new services in the future.

The trigger for the Hometown upgrade is to increase channel capacity, and it can be demonstrated that fiber optics "prove in" economically versus competing technology (in particular an AM microwave radio link - a system commonly known as "AML" radio) in trunk applications. This would equate to a $0 increment to be spent on new technology; i.e. we pay no more to get into fiber, but the $0 increment may not provide a solution which meets all of the strategic objectives.

In 1988 the example of a channel upgrade was analyzed and using industry financial benchmarks the prove-in points for various technology options were found. One interesting side effect of this approach is illustrated by the graph of investment for fiber optic options. This graph plots (on a log-log scale) the incremental revenue per subscriber versus the incremental capital cost for the various fiber optic options over and above the $0 channel upgrade reference scenario. The conclusion of the study was that for $1 or less per month of revenue, a CATV system could make a valuable investment in fiber optics. This equated to about $40 per subscriber in incremental capital. The issues then are:

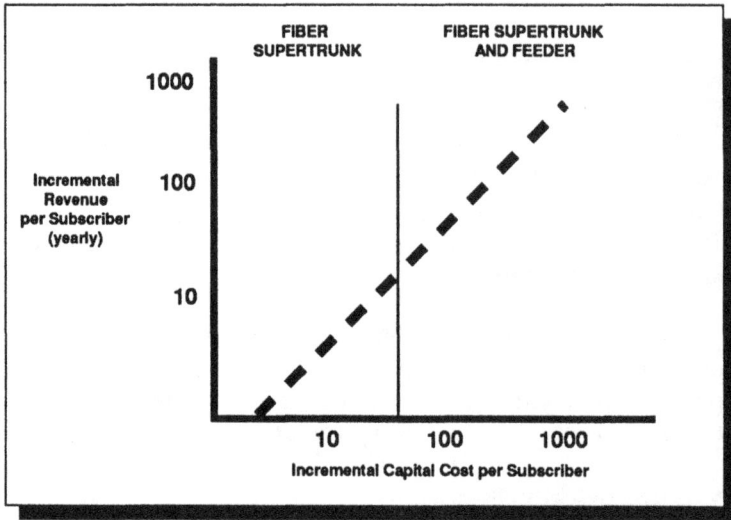

Graph of Investment for Fiber Optic Options

• How much incremental revenue should I expect per subscriber because I deployed fiber optics?

• How do I decide between the various fiber optic options? (e.g. the $4 per subscriber in capital cost can be spent either on a digital fiber supertrunk, or on an analog supertrunk and feeder system).

Step 2 - Investment per Subscriber

A benchmark $100 capital investment per subscriber was presented.The level of investment desired will be highly system dependent and probably subjective. It is, however, advisable at the outset to pick a level of investment as a target that is somewhat higher than the $0 breakeven point for the channel upgrade. This has two effects:

• it allows for greater choice in making technology selections;
• it provides greater flexibility in evaluating architecture options.

These effects can have a very big impact on the deployment plans, particularly on reliability and futures objectives. Implementations can be hypothesized and then taken into account, but it is better to have an idea at the outset of what the tolerable expenditure level should be. Using the iterative approach outlined in Chapter 1, we are able to make successive modifications to the plan to knock the cost down. We want to find a minimum cost implementation which meets the strategic objectives, i.e. it will likely not be the absolutely lowest cost solution, but it will be the lowest cost solution to do what we want to do.

NOTE: $100 is not our"recommended" investment level; a system operator should choose his own number and then use the methodology to assist in the decision making process. Certainly not all the interviewees agreed with $100 per subscriber as a target.

Step 3 - Budgeting the Investment Objective

The next step for Hometown was to budget the $100 per subscriber over the various objectives that make up the strategy. The following breakdown for how to spend it was proposed in the survey:

• 85% for signal quality;
• 10% for system reliability;
• 5% for future-proofing.

In general it was felt that reliability should have a greater weighting and that the futures budget was too low. The system operators felt they should invest more in a competitive infrastructure for new services.

These comments led to the following revised budget:

• 45% for signal quality;
• 40% for reliability;
• 15% for future-proofing.

NOTE: These are also not to be taken as our recommendations. They are benchmarks to be used to develop the Guide methodology. Each system operator will be able to make his own budget and then plan for the new technology accordingly.

Step 4 - Defining the Technical Parameters

In parallel with the investment level decisions per subscriber, certain technical parameters were defined which reflect the strategic objectives.

Set-top Signal Quality

The interviewees believed that they need to be ready for HDTV, and that set-top signal-to-noise ratios (SNR) will have to be in the order of 49-50 dB. While fiber optic technology can improve the set-top SNR, the power of the technology is not in achieving a particular level, but in achieving a consistent level throughout the system. The approach in Hometown is to pick a level for set-top quality, and then budget the noise parameter through the system, working up to the head-end.

A 49 dB signal-to-noise ratio objective at the set-top was initally proposed. The CATV system would then only be allowed to add a given amount of noise for any subscriber. The objective was felt to be conservative; a figure of 50 - 52 dB was felt to be a better target.

The Hometown strategy is to assume 50 dB at the set-top. This objective will be used to drive the specification of other points in the system. These numbers could be varied from system to system depending on the operators specific objectives [3].

System Reliability

An overall figure of 99.997% for system reliability (not including power-related failures) was proposed in the survey. This figure is from a Telco that had been considering deployment of fiber optics for CATV type services [4]. The figure is aggressive (15 minutes per year of downtime) but does represent the type of reliability customers have come to expect from the telephone system. This was the first objective of Hometown.

It was difficult to derive a representative number for the existing systems due to the large skewing of reliability due to dependency on ac power. (Investment in power back-up improvement should perhaps be considered in conjunction with fiber optics.)

A figure of 99.8% was suggested as representative of the existing systems. An improvement to 1.5 to 2 hours per year would provide an objective of 99.98%. If amplifiers are considered to fail at 2% per year and cable and power failures are not included, the objective of 99.98% would equate to a cascade of about 20 amplifiers (depending on the mean-time to repair objective chosen).

The figure was modified to the objective of 99.98% availability. This should be reviewed for particular systems and in light of the competetive infrastructure.

Future Proofing

In Hometown, we are mainly considering near-term implementation objectives. The future-proofing parameter represents an investment a system operator would be willing to make now, to ensure long-term viability. The most obvious investment to make is in selection of an appropriate number of fibers per cable route to allow for possible new services, and in ensuring that the physical topology of the system does not freeze a specific architecture. Fiber optic technology is still very much in its infancy; future technologies may very much change the nature of the CATV system. Electronics is assumed to be a five-year investment, but fiber cable should be considered as a 20-30 year investment.

The opinion among the interviewees was that, for near-term deployment, the strategic element was to size the fiber optic cable appropriately and to spend the futures budget on that parameter.

2.3 Developing the Plan

We now have a top-down budget defined, and it is necessary to allocate the investment and technical parameters over the different parts of the CATV system. The split of plant for a typical system (assuming all coaxial cable) was:

- trunk   15%
- feeder and hub   40%
- distribution   45% [5].

To differentiate feeder and hub from distribution is often difficult. The above bench-marks are only important to help to weight the parameters over the parts of the system. They give us an indication of how much we should expect to spend. (For example, for a given territory, the supertrunk carries 100% of the subscribers and should take a proportionally greater portion of the expected investment than a feeder link which typically may carry 10%.)

The following weightings will be used:

	% of Subs	% of Plant	Weighting	
Trunk	100	15	15	(75%)
Feeder and Hub	10	40	4	(20%)
Distribution	2.5	45	1.1	(5%)

Signal Quality Weighting

It is very difficult to weight the signal quality improvement without fixing the number of amplifiers in the end section of the system and without strictly fixing a topology. The 50 dB figure at the set top is assumed to be roughly a 7 dB improvement over the present level. If this improvement is applied linearly over the system, the supertrunk would have to meet 59 dB typically (assuming a 30 amplifier cascade, and 15% of the plant in the trunk). Since the trunk portion of the system is weighted strategically higher than the others, the expected improvement should also be higher. Linear application of the improvement also tends to fix the number of amplifiers allowed in the distribution plant.

Somewhat arbitrarily, the trunk objective has been set at 60 dB and the feeder-hub objective at 55 dB. Fixing these points allows for flexible system topologies, provided the other objectives for cost and reliability can be met with available technology.

Reliability Weighting

The weightings used for reliability use the "number of subscriber" figures only. For any given link to the subscriber, the trunk is weighted 100, the feeder 10, and the distribution 2.5 to reflect the relative importance of each segment. For reliability calculations, the weightings assume that the trunk should be eight times better than the rest (i.e. 100/(10+2.5)), and the feeder four times the distribution. This sets up a set of three simple simultaneous equations to calculate the number of allowable hours of outage per segment, given a system operator defined objective for annual hours of outage.

In Hometown, the objective is 2 hours of outage time, which means we allow 0.22 hours for the supertrunk, 0.36 hours for the feeder and hub, and 1.42 hours for the distribution portions of the plant.

2.4 Hometown Cablesystem Deployment Benchmarks

Using a benchmark of $100 per subscriber invested, the following plan has been developed for Hometown Cablesystem:

Signal Quality:

Invest $45 per subscriber to get a 50 dB signal-to-noise ratio at the set-top.

		Investment/sub	SNR Objective
Per Segment:	Trunk:	$33.54	60 dB
	Feeder and Hub:	$ 8.95	55 dB
	Distribution:	$ 2.52	50 dB

Reliability:

Invest $40 per subscriber to get 2 hours per year of outage time.

		Investment/sub	Outage Objective
Per Segment:	Trunk:	$29.81	0.22 hrs
	Feeder and Hub:	$ 7.95	0.36 hrs
	Distribution:	$ 2.24	1.42 hrs

Future Services-Bidirectionality:

Invest $15 per subscriber to ensure architectural and infrastructural longevity.

		Investment/sub
Per Segment:	Supertrunk:	$11.18
	Feeder and Hub:	$ 2.98
	Distribution:	$ 0.84

This set of parameters will be used to evaluate the various fiber optic technologies and products available for Hometown Cablesystem. Implementing fiber optic systems which satisfy these parameters would ensure that the system's strategic objectives are met.

The numbers provide us a guideline to help with the implementation decisions needed later. We will not necessarily use them strictly, but as an indicator. The main goal is to meet the strategic objectives for less than $100 per subscriber. The split of this over the system and the technical parameters indicates where we should focus on spending the money. The technologies available will determine how much it will actually cost.

References

1. *International Journal of Digital and Analog Cabled Systems*, Vol.1, 55 (1988).

2. *The Wall Street Journal*, February 23, 1989.

3. Noise is a complicated subject in television systems. We have simplified it somewhat here by only considering signal-to-noise ratio. In a real system we should broaden that specification to include the second and third order of harmonic distortions. In a noise budgeting process these should be defined for the various points in the CATV system as well; i.e. we would attach several noise specifications to the 50 dB at the set-top requirement, for example, 65 dB composite second order and 65 dB composite triple beat. Noise is discussed in more technical terms in Chapter 4.

4. Private source.

5. In some ways this assumes an architecture, but this does not limit the analysis to any particular solution.

Chapter 3
Architectural Considerations

3.1 Long-term Architectures for CATV

What is an Architecture and Why do I Need to Know About it?

Cable television has always had an architecture. An architecture according to our copy of the *Oxford Dictionary* is a "style of construction". The style of construction used in CATV is known as "tree and branch" and it was essentially dictated by the broadcast nature of the service and the coaxial technology used. The plant is designed and optimized for broadcast distribution. The architecture is known as tree and branch, reflecting the main trunk line concept and the successive splitting of the broadcasted signals out to each subscriber.

Tree and Branch Architecture

Tree and branch architecture fully meets the needs of providing broadcast type distribution and it is inexpensive. It does have its limitations, the principle ones being:

• It is difficult to implement anything except broadcast services.

Markets a CATV operator may want addressed over and above broadcasting are, for example: narrowcasting of services to specific neighborhoods, or bidirectional data services for large businesses. While technically possible to provide these services on a tree and branch network, the numerous approaches to doing so have never really taken hold.

• It is susceptible to outages for large numbers of subscribers.

This is because each large branch of the system is broadcasting to very significant portions of the total base of subscribers. It is a "shared facility". A single amplifier failure in a CATV system effects at least several hundred subscribers, and in the case of a supertrunk, perhaps 100,000.

- <u>Performance is inconsistent throughout the system</u>.

In tree and branch, every subscriber "sees" all the way back to the head-end. Those close to the head-end may see 8 or 10 amplifiers, those further away may see 40 or 50. Thus, subscribers practically by definition will receive different levels of signal quality and reliability depending on their location in the system.

- <u>It is difficult to identify and sectionalize system failures</u>.

Since the information flow in the system is unidirectional and away from the headend, the CATV system operator only knows there has been a failure when the subscriber calls in to tell him.

Despite these limitations, the CATV business has done very well by the existing tree and branch approach. There is no question of throwing it all out or treating it like it is a negative asset; there is still simply no more efficient nor more economical way to distribute large numbers of broadcast television channels to subscribers' homes.

There is however, a new opportunity presented by fiber optics. That is: to build a new style of system avoiding the limitations introduced by tree and branch and building in new capabilities. The economics of CATV distribution are that the "last mile" of coaxial cable to the subscriber will be in place for many years to come (as will the copper pair for the telephone). The question for the system operator today is how to strategically deploy fiber optics to reduce the amount of tree and branch in the system. A CATV operator should plan to evolve the system making successive improvements to the architecture as the fiber technology matures.

Terminology

There is a mixing of names in the industry which causes some confusion in discussing architectures. For our purposes we will consider a "trunk" to be a transmission line from a head-end to a hub location, or between head-ends, or between hubs. "Distribution" will be from the hub location to the subscriber. Using this definition, an existing system with all coaxial line would be all distribution, which is consistent with popular terminology [1].

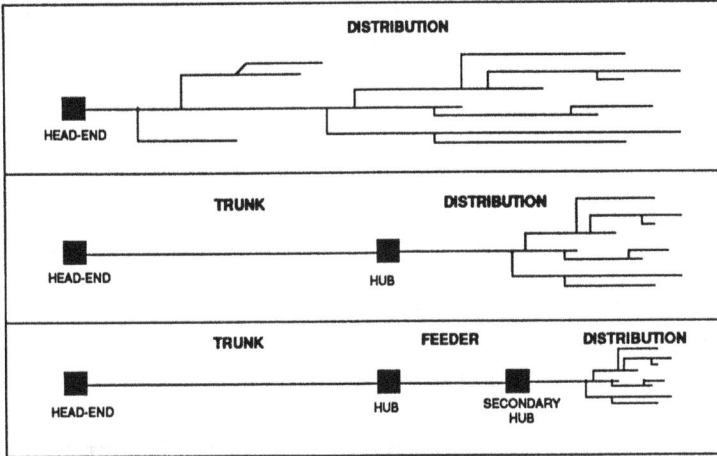

Link Terminology

Many systems have implemented AM microwave radio trunk systems so the distinction of trunk and distribution is already there. For our purposes this also allows us to consider comparisons of fiber systems versus radio for trunks and fiber systems versus coaxial for distribution. We also introduce into the system a "secondary hub" (sometimes called a "subhub") which uses fiber to connect back to the hub. This link will be called the "feeder" and in a traditional system it may have used coaxial trunk amplifiers. This causes some confusion, but it is technically accurate as the feeder plant is built like a secondary trunk and looks much the same.

Fiber Optic Architectures

In theory one could build a CATV system based on a fiber optic tree and branch architecture. Since the same limitations would conceivably (but not necessarily) exist as in today's coaxial system this is probably not a good architecture to pursue. In addition, technologies to implement a fiber tree and branch system (for example: low-loss splitters, optical amplifiers) are not yet commercially available.

From a planning point of view, the objective should be to try to minimize the amount of coaxial tree and branch plant in the system. In the context of the methodology outlined in Chapters 1 and 2 this means: implement as much fiber as possible in order to meet objectives while staying within the overall budget. But what architecture should be considered of the three basic types of telecommunications architectures: tree and branch, star, or bus?

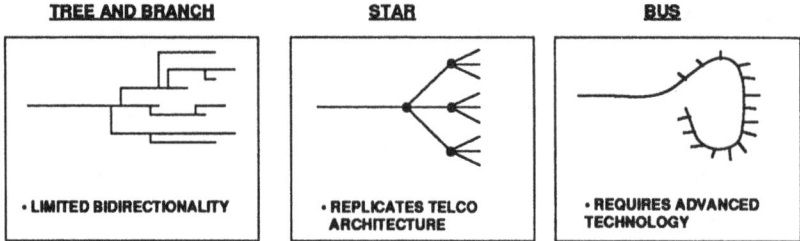

Types of Architecture

Star architecture is what the telephone company has. It is optimized for bidirectional services, and when implemented with copper pairs is the most efficient and economical way to provide telephone service. When considering broadcast services, this architecture starts to look very expensive compared to the tree and branch approach of CATV. Copper wiring for the telephone plant generally runs around US$1000 per home [2], versus CATV which is less than US$500 [3]. To do CATV distribution using fiber in a star type architecture using today's technology (i.e. commercially available systems) would run in the range of US$2,000-$4,000 per home [4]. No one actually agrees on any particular figure. While a study by the National Cable Television Association claims it to be possibly as high as $5,000 to $10,000 per subscriber, meaning a total $450-$900 billion to rewire the U.S. [5], Bellcore obtains a total of $200-250 billion, and justifies the investment in the enlargement of the revenue base generated by the new services [6]. But without very large additional revenue streams, this architecture for fiber to the home will remain prohibitively expensive for some time.

Star architecture however does offer some very significant advantages relative to tree and branch and solves most of its limitations. In particular for trunks and feeders, the star approach can be proven-in economically and can be used strategically to help address signal quality and reliability objectives that may exist for the system.

Bus architecture is similar to that of the computer communications world. This can be viewed as being similar to a garden hose running down the street. Everywhere a hole is punched in the hose, the water comes out [7]. At some levels, the bus approach is similar to tree and branch, in that it provides a shared facility over a number of subscribers, thereby reducing costs. A bus system would be designed with a "read bus" and a "write bus" thereby providing bidirectionality, and would also have identifiers built in for each subscriber connected to it. In this way a bus system can overcome the limitations of tree and branch.

However, if a bus system were used all the way from the head-end to the subscriber, it would also suffer some of the disadvantages of tree and branch (example, it would be difficult to provide the same signal level throughout the system; there would be a lot of subscribers on the same facility, etc.) and it is probably best viewed as a distribution technology; i.e. to go the last mile to the home.

The bus approach for fiber is very good, given that the components come available to make it economical to implement. BellCore has published results of a study showing a bus type architecture running at between $600 and $2,000 per home [8]. Raynet Corporation has published a proposed "star-bus network" which holds promise in this area [9].

Once one of these topologies is chosen, cost limitations will likely induce some facilities sharing among a certain number of subscribers. The ideal size of these clusters remains to be calculated.

As a common point, the affordability of these topologies involves components such as the evolution of the capacities and prices of the optoelectronics, especially optical switches and optical interfaces in the homes, or as the evolution of techniques and prices of the passive elements: e.g. the fiber itself, the splicing techniques, the coupling elements (price drops are announced for a near future [10]).

The ultimate fiber architecture is still open to question. The "holy grail" is fiber-to-the-home, and there are many stakes in the ground and a lot more to come before it is all sorted out. Our prediction would probably lean toward a combination of star and bus to get the best of both topologies.

3.2 Trunking and Distribution Considerations

In Chapter 2 the system was divided into conceptual pieces in order to apply financial benchmarks and technical budgets to it. Considering trunking and distribution individually is important architecturally as well, particularily when one addresses reliability. The trunk by definition carries all the subscribers served by that trunk. As we consider distribution the relative number of subscribers per transmission link decreases rapidly the closer we are to the subscriber. Thus, from a system design point of view, this is considered against the type of performance expected from each link; i.e. a single link with 100,000 subscribers should have a much greater reliability than one with 10,000.

There are two basic system types against which the system architecture is planned:

• If the system already has an AM microwave trunk.

In this case we have already got a start on the architecture because we have a hub point identified. If we do not want to move that hub point (for financial reasons, or for convenience,etc.) then our equation for the fiber architecture is simply seeing if we can get a point-to-point fiber system that better meets the objectives we set for the trunk portion of the plant; i.e. it is a basic technology selection problem.

• If the system is all coaxial plant (or if one is willing to move an existing hub).

In this case the placement of the first hub may not be obvious. The first step in architecture is to find where that hub goes so the plant can be split into trunk and distribution. Using the objectives from Chapter 2 we should find that point by choosing the technology which meets the channel capacity and signal quality objectives, while staying within the allocated dollar budget. This will be the first cut at locating the hub, which can be optimized on subsequent iterations.

For very large systems we may need to review the distribution of population over the area to determine the approximate number of trunk lines needed to establish the first hubs. A first cut at an architecture for a large system would be to center the hubs based on population density, then apply the trunk objectives from Chapter 2 to each one. The exact placement and number can be optimized on subsequent iterations.

The placement of these first hubs and the optimization of the technology to serve them from the head-end is relatively straightforward. A bigger challenge for the fiber optics technology is to move economically to the next step into the feeder; i.e. to serve a set of secondary hubs.

Secondary Hubs

The approach is to initially get an idea of how many secondary hubs are needed and where they should go. There have been a number of methods for choosing secondary hub locations. You should optimize the secondary hubs based on the strategic objectives defined in Chapter 2.

The architectural motivation is to minimize the amount of coaxial tree and branch plant left in the system. This concept is illustrated in the figure below. This figure is highly simplified to illustrate the technique. As we get closer to the subscriber (i.e. moving to the right along the horizontal axis) the amount of fiber optic investment one gets increases while the amount of coaxial plant remaining decreases. The objective is to find the point at which they cross, which defines the minimum cost implementation to meet the objectives. We will find a point using this technique which is not necessarily the lowest cost point, but one which maximizes the amount of fiber we can "afford" based on our strategy. The curves are actually considerably nonlinear as you will see in the Hometown Cablesystem section of this chapter which very quickly limits how far you can take fiber using today's technology (unless you have a very aggressive strategy). The important aspect of the analysis though is that you can apply it as technologies or strategies change, and find new optimal points under those conditions.

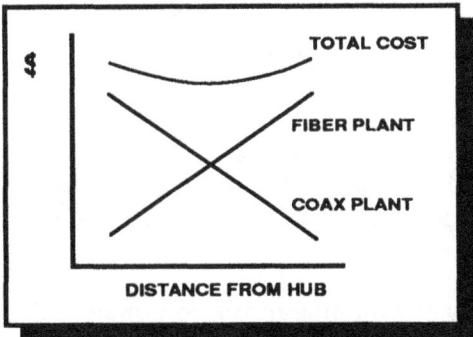

Cost of Hybrid Plant

To reassure oneself of the result of this analysis, one would be prudent to look at the secondary hub to subscriber portion, the remaining distribution plant, to ensure that the objectives for that are still met. Based on the budgeting technique employed in Chapter 2 it should meet or exceed the objectives.

3.3 Reliability and Bidirectionality

Once a basic architecture has been established to meet channel capacity and signal quality criteria, the system should be reviewed against the reliability and futures objectives.

By pushing the fiber optic technology as far as is affordable into the system, a certain amount of reliability improvement and built-in futures will already be apparent. At this point the system needs to be tested to see if it has been modified enough to meet the objectives.

Reliability

Fiber optics can improve reliability in three ways:

• By replacing active serial elements (i.e. coaxial amplifiers).

A fiber optic system essentially has a transmitter and a receiver which both contain active components. In between, which can run anywhere from 5 to 100 km depending on the system being selected, it is passive. Hence there is a very significant reduction in system outages just by placing fiber instead of using cascaded coaxial amplifiers.

• Through providing enhanced performance monitoring and protection switching.

Even though the fiber cable is passive, the electronics on the end of the fibers can fail. In particular, the component most likely to fail is the laser. In most telephone company systems this eventuality is protected against with a protection switching device at the high speed electrical level, built-in just before the laser.

• By providing an alternate route.

This will reduce the amount of cable related failures. (In the trunk plant there are obviously a number of other technologies which can provide alternate route facilities, such as microwave radio.) In the telephone trunk plant it is often relatively easy to implement an alternate route by putting the fiber carrying the standby protection system in the diverse cable route.

Generally in fiber optic systems designed for CATV today, these latter two approaches to reliability are not being addressed. It is important to analyze the system against the reliability objectives to ensure they are being met.

Reliability does not come for free.

Calculating System Reliability

To get an idea of your system reliability you need two important parameters: failure rate of the equipment you are using and the time it takes to repair it if it fails. The vendors of fiber optic electronics for CATV are sometimes reluctant to give out figures for this as the technology is still new (particularily for the AM systems). By considering the critical element to be the reduction in coaxial amplifier cascades, you can get an idea of your system reliability as follows:

Example:

Assume a 40-amplifier cascade where amplifiers fail at 2% per year [11]. The failure rate will therefore be 0.8 amplifiers per year for the subscribers at 40 amplifiers away. If it takes you on average seven hours to fix the amplifier, your yearly expected downtime will be 5.6 hours; your system will be available for 99.93% of the year (based on an 8760 hour year). This gives a kind of best case view for the coaxial plant as it does not include cable failures or power outages (as a reference, your telephone is working on a network designed to meet about 20 minutes or less per year; i.e. it is expected to be available for about 99.997% of the time [12]).

Without knowing the failure rate of the fiber optic equipment, a representative number can be found by considering the fiber system to be the equivalent of one amplifier (with a lot more cable in between).

It is important to insist on receiving failure rate information from the vendor even if it is only a theoretical calculation. Bell Communications Research [13] among others have published techniques for doing this.

Bidirectionality and New Services

Once you have established your system architecture and your base technology selection you can apply the budget to build in futures into the system. By pushing the fiber as deep into the system as is affordable, we have built in some future capability. Additionally though we can look at the budget we started with and look at some things, such as:

• Adding extra fibers.

This is the best way to build in futures for just about any contingency. An algorithm for sizing the fiber cable is presented later in the next section.

• Changing hub locations.

Depending on your level of aggressiveness you can use the futures budget to increase the number of secondary hubs to get fiber closer to the subscriber, or perhaps move the hubs around to provide better access to potential new customers (such as large businesses or universities).

• Overlash fiber cables on the coaxial plant.

For the sections of plant where fiber was not needed to meet the quality and reliability objectives, you may consider adding some fibers just as a future investment. This decision will need to be balanced against projected drops in fiber pricing versus the convenience of placing it now during an already planned rebuilding. What is done with the "futures" budget is highly subjective. Most system operators we talked to were fairly conservative in wanting to allocate money for futures. Based on this, the best approach is to add extra fibers where you can. This is a future that can be built in at a relatively low cost.

3.4 Outside Plant Planning

The strategic element of the outside plant plan for fiber optics in CATV is to get the fiber count right. The other elements of outside plant are much the same as for coaxial cable, in terms of designing a cable route whether it be aerial, buried or in ducts.

Two important elements that require consideration at the planning level are as follows:

• splice locations.

Fiber cables come on various reel lengths. Cable routes should be looked at to ensure that splice locations are at desirable places [14].

• number of splices.

Minimizing the number of splices is a good objective to conserve the loss budget of the system, but should not be a critical factor.

Fiber Count

In determining optimal fiber count, the following assumptions were made:

• plan to be optimized for a five-year horizon.

We expect the fibers themselves to last "forever" [15], but our planning view can only stretch five years.

• investment to be paid back in five years.

The equipment should be paid back through revenues in five years or less. The useful life of the equipment should be much longer, but five-year pay back gives a conservative financial view.

The sum of these two assumptions is that after five years if our objectives have changed radically, the underlying technology will also have changed. The fiber count will still be good.

For example: Building a 60 channel system today using eight fibers. Within five years I do not expect to have to go to a higher number of channels. After five years if I want 150 channels, I can likely get equipment to provide that using the same eight fibers or less. The rapid historical pace of fiber optics technology would tend to support this. If I expect to have to go to 150 channels within the five-year horizon, I should reconsider my fiber count or my technology selection.

Determining Optimal Fiber Count

Asking around for some advice regarding the optimal number of fibers you should implement in cable links, you may very well face all sorts of answers, from "six !" to "as many as you can buy !".

The fiber count determination for a system is an architectural issue. It depends somewhat on decisions taken at the technology level of the planning and thus would normally take place in a second iteration of the strategic plan.

One important thing to keep in mind here is the fiber cable economics: the price of a fiber cable is not proportional to the number of enclosed fibers. The cost of its installation through a contractor is even less proportional. And having a contractor come back after three years to pull another cable along the same path can be very expensive.

Example:

One type of cable priced at $3.20 per meter with four fibers ($0.80 per fiber) will be around $5.80 with 12 fibers ($0.48 per fiber), and $29 with 72 fibers ($0.40 per fiber). A contractor will ask the same price to place a 4-, 6-, up to 12-fiber cable, for example $2.00 per meter; this will become $3.00 per meter for a 16-fiber cable [16].

The problem for the system planner will often be to quantify the qualitative. Here are some guidelines to a systematic approach of the different contributors to optical fiber count in a link.

• Number of fibers needed, based on the number of channels:

First and easy step, the basic number of fibers we want to implement derives from the number of channels to be transmitted for the current requirements, and the signal quality to be achieved.

From the first iteration through the planning process we will have determined basic technology selection, and the maximum number of transmitted channels per fiber in order to achieve the signal quality we have set up as a strategic objective, (note that the dynamic process of system planning may conduct us further on to reconsider our technological choices; the number of transmitted channels per fiber may vary as a consequence, and thus the number of fibers needed).

This figure known, we must consider that if we just deploy a cable with five fibers, we may fail to meet some of the strategic objectives, i.e. we may run out of fibers within our five-year plan, even though the simple fact of rounding up the number of fibers ordinarily provides the system with some idle space. In practice the cable may have to be rounded further up if the fiber modularity is not available.

It is natural to admit that some additional requirements in terms of number of channels are likely to arise within the duration of our plan, for which we should provide fibers. This depends on the corporate strategy in that matter, and the evolution of the outside world.

Additionally, two spare fibers should be provided for eventual maintenance actions requirements, described in Chapter 5.

Example:

To transmit 58 channels from a downtown head-end to a promising area being urbanized 20 km away, we decided upon a FM system guaranteed to transmit 12 channels per fiber and meet an objective of a 60 dB signal quality. Therefore we need five fibers.

Let us say that, in an adjacent region of operation, competition is expected to provide more channels. An increase of 98 channels has been determined at year three for a total of 156. To cope with this increase without having to get the cable installation teams back in the field, we would need to add eight fibers to the cable, if we keep the same specifications of 12 channels per fiber.

This tends to produce a "worst case" result: planning tomorrow with the technology of today. Manufacturers are aiming at the transmission of more and more channels per fiber, and some improvement can be expected within the duration of our study. A careful review of the plausible progress may lower the channel growth factor, though delays for a breakthrough are difficult to quantify.

• Reliability and survivability considerations:

Reliability and survivability are part of the strategic objectives. They are reflected in architecture decisions which are here translated into numbers for fiber count.

Link protection [17]:

If the link was foreseen to be protected 1 to N, the number of fibers in the cable is multiplied by (N+1). (That was easy!)

Alternate routing:

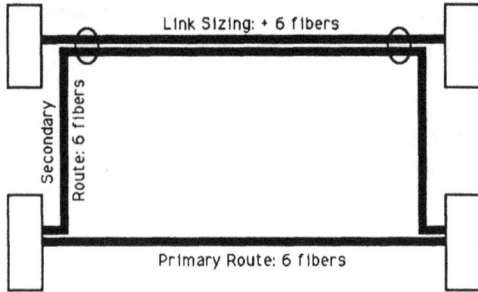

Link Sizing: + 6 fibers

Secondary Route: 6 fibers

Primary Route: 6 fibers

If the link under study supports an alternate route for protection against cable cuts the number of fibers calculated on these links are added to the cable size, and the additional cost is taken out of the primary route links budget.

Alternate Routing

- **Planning for the future:**

Protecting future, whenever possible, is another strategic objective, and was attributed a budget when we set the economic benchmarks. This budget is to be divided between electronics and fibers.

The share between electronics and fiber costs in a link varies according to the technology and the service (for example, bidirectional data service requires transmit and receive functions), among others. As a basis, fiber cable can be estimated to represent one fifth of the cost of a link providing a high quality service, and much more, up to one third, for a distribution link (a quick review of different vendors in the technology chosen would be useful, at planning time, to derive a more precise figure).

The fiber cost percentage found, e.g. 20%, is then applied to the budget for the future to obtain the best amount of money to be spent on extra fibers in the cable.

The number of extra fibers we can afford for future insurance is then derived from manufacturers and installation contractors prices. To be effective, this number should not be lower than four to provision for a "data" type service (a data type service is bidirectional and protected).

• Shared cables:

Some cable portions may be shared by two different destinations following architecture decisions previously made. There are two different cases:

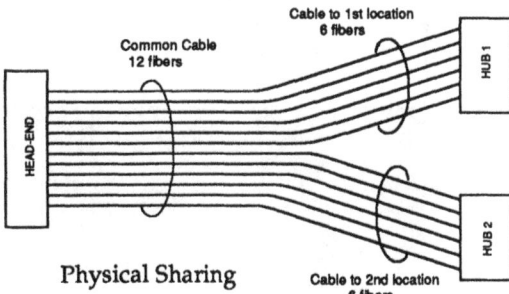

Physical Sharing

In the case of **physical sharing** fibers dedicated to different locations physically share the same sheath on a certain distance, given that a cable containing $2 \times N$ fibers will be much cheaper than two cables containing N fibers. Hence we simply sum the number of fibers calculated for each link to obtain the definitive cable size.

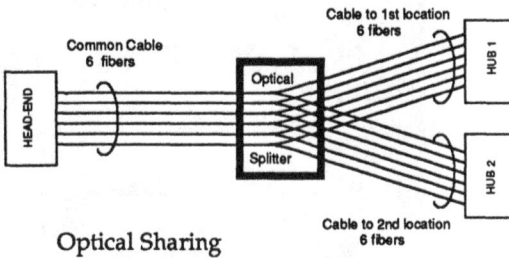

Optical Sharing

In the case of **optical sharing** the information carried by a single cable is split into two or more identical cables to service different locations; resulting in fiber saving on the common section. In the total budget, care must be taken for the common section to be counted only once, and the price shared between the different links serviced.

• Possibility of leasing:

Knowing the common interest in fiber optics for private network, a market study of the leasing possibilities for optical links could identify new revenues.

It would then be a corporate decision to consider leasing as a help to the justification for the system. Two options are to be considered: lease of idle facilities until they are required for growth, or adding fibers to the cable for more permanent leasing.

The different contributions to the optical fiber count can be summed up in a Strategic Fiber Count Form, as illustrated.

STRATEGIC FIBER COUNT FORM	LINK DESIGNATION
	Tx End: _____
	Rx End: _____
	Number: _____

Basic number of fibers

Reliability and survivability factor +

Future +

Foreign fibers sharing the cable +

Optically shared fibers −

Leasing +

Cable size

Fiber Count Form

3.5 A Review of Some Architectures

A number of articles have been published recently on architectures for fiber optics in CATV. Some of these are reviewed in this section. These examples should be used to generate ideas and alternatives for the reader's own system. Selection of an architecture is system dependent and should be driven by the strategic objectives set for the business.

The ATC Fiber Backbone [18]

American Television and Communications Corporation (ATC) has published a number of papers on their approach to implementing a fiber backbone. The approach considered the following present capacity typical of ATC systems:

Trunks	50-300 MHz	(35 channels),
Distribution	50-300 MHz	(35 channels),
Drop	50-1000 MHz	(150 channels).

The objective of the ATC approach is to unlock the "drop" capacity by improving the bandwidth characteristics of the trunk and distribution plant; i.e. to get 150 channel capability to the drop. ATC set the immediate objective to be to provide 80 channels in the backbone and to reduce the amplifier cascade to a maximum of four; i.e. you do not quite get to the drop point in the system. This would solve a 10-year capacity problem.

The backbone approach breaks up the system into dozens of small cable systems. The backbone provides a number of strategic benefits including reliability, signal quality, channel capacity, simplicity, flexibility, and competitiveness (i.e. it provides a strategic architecture on which to build the business). The implementation cost is estimated at $30 - $60 per subscriber depending on which of the benefits you choose to exploit. This investment level is felt to provide the 80% improvement level; i.e. to push the fiber closer to the home would not yield significant added benefit.

Based on this architecture ATC also publicized their specifications for a fiber transport system:

60 Channels,
55 dB Carrier-to-Noise Ratio,
65 dB Composite Triple Beat,
65 dB Composite Second Order,
65 dB Crossmodulation Products,
10 dB Power Budget.

(Note to the reader: for a discussion of the decibels, turn to Chapter 4.)

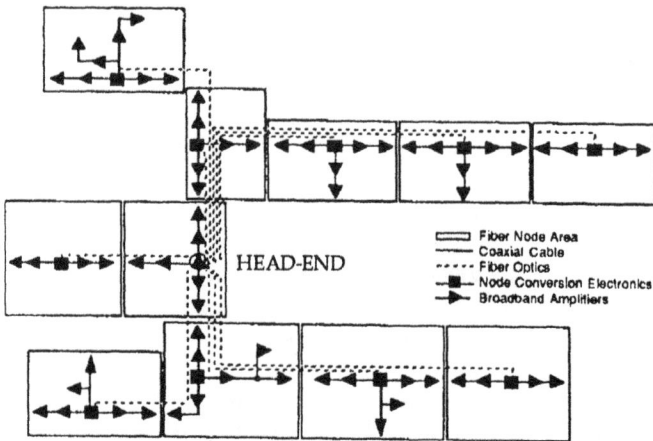

ATC Fiber Backbone Approach

(Source: Chiddix, J., op.cit.)

The key elements for the CATV operator of this approach are:

• good investment profile ($30-$60 per subscriber),
• reduces amplifier cascade to a maximum of four,
• uses all of the existing amplifiers by turning some around to serve subscribers ahead of a hub point,
• currently provides no redundancy in the fiber routes (although this is being explored in a multihub system in Orlando, Florida).

Jones's Cable Area Network [19]

The Jones model is very similar to ATC, except that the existing plant is left as is and used as a redundant path in case of fiber failure. It is simple to implement during a rebuild or upgrade due to the minimum rearrangement of the existing coaxial plant. Additionally it can be argued that the maintenance cost of the system would be the same as the ATC approach because there are still the same number of amplifiers.

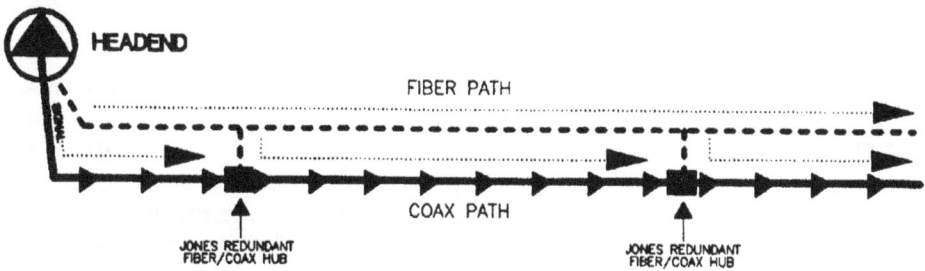

Jones Fiber Configuration

(Source: Luff, R.A., op. cit.)

Cautionary Notes:

• Even though the cascade is reduced to a maximum of five amplifiers, the redundant path is coaxial all the way from the head-end. In the future, if the system is to have 150 channels, some of which may be HDTV channels then the engineering of the 40 amp cascade in the redundant route may be somewhat different from the 5 amp cascade in the normal route.

• From a reliability viewpoint, the redundant route will likely be worse than the fiber route. Thus, in all probability, if there is a failure somewhere in the fiber system, there will already have been a failure somewhere in the coaxial system. This raises the whole issue of remote monitoring, on coaxial tree and branch systems which would have to be solved for the redundancy scheme to work.

This is not to say that one should not consider the Cable Area Network (CAN) approach, only that when looking at various architectures be sure not to lock out some of the strategic objectives you may have. The CAN technique may be somewhat less expensive than the ATC backbone providing it meets your business objectives.

The key elements of the approach are:

• very similar to ATC, perhaps less expensive,
• suggested reduction to five amplifier cascade maximum,
• leaves existing coaxial plant intact,
• routes fiber down a neighboring street to provide some cable diversity,
• may have some hidden traps from a strategic perspective.

Rogers Fiber Architecture [20]

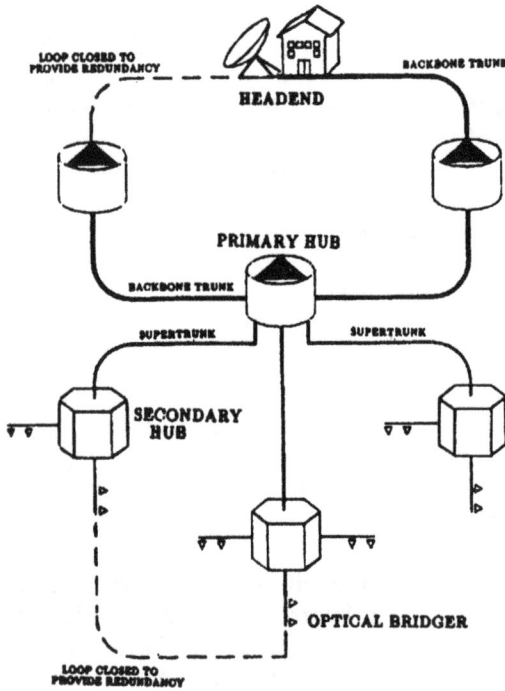

The Rogers approach is also similar to that of ATC with the addition of redundant links in both the trunk portion of the plant as well as from the hub points out to the secondary hub or "bridger" locations. Studio quality is maintained at the hub points by using FM transmission from the head-end to the primary hub. The distribution plant is designed so that each bridger amplifier location is served by its own set of fibers.

This architecture will be somewhat more expensive than that of ATC due to the extensive use of cable diversity and will probably have larger fiber counts.

Rogers Fiber Architecture

(Source: Hart., G., *Rogers Fiber Architecture*,
CCTA Technical Papers 1989, p.52)

SaskTel Linear Photonics [21]

SaskTel is the provincially owned telephone company in Saskatchewan, Canada. In 1977 it was granted the exclusive right to carry and deliver cable television services in the province. In order to serve the large number of geographically dispersed communities, SaskTel constructed a fiber optic digital broadband network which was completed in 1984.

Linear photonics is a 1988 development which provided for transmission of 16 TV channels over fiber. The fiber was run in parallel with the coaxial plant which remained in service carrying the original 16 channels. In this way SaskTel carries 32 TV channels split between the coaxial and the fiber. The 16 fiber-carried channels are AM modulated and block converted into an optical signal. This signal is injected into the coaxial plant at a trunk bridger amplifier for final transmission to the subscriber.

While being another variant on the ATC backbone approach, the SaskTel architecture will likely have costs similar to the Jones Intercable architecture and would be comparable to adding incremental channels on the coaxial plant. In addition, having the 32 channels split between the coaxial and the fiber means that for any facility failure only half the channels are lost.

Other Architectures

There are a great number of fiber to the home trials underway in the telephone industry (21 are quoted in Kaiser [22]). These promise to shake out architectures and drive the technology. Unfortunately for CATV operators, these trials are generally trying to implement bidirectional services (i.e. telephony) as a high priority. Their cost limitations are also driven by the need for active optical components in the home. For example, the Alcatel system being tried in Bell of Pennsylvania has subscriber termination costs for CATV of roughly US$3200 for only the subscriber end portion [23]. The BellSouth trials in Florida have been estimated at over $10,000 per home [24]. These types of costs have eliminated most of the approaches from consideration for CATV applications for the present time.

A number of proposals to take fiber closer to the home using advanced fiber optic techniques exist such as splitters or wavelength division multiplexing. Architecturally these tend to be limited by the cost for the "last mile" portion of the plant, that closest to the subscriber.

<u>Southern Bell in Heathrow</u>

Southern Bell's trial in Heathrow, Florida, implements an entirely digital approach to fiber to the home developed by Northern Telecom and Bell Northern Research [25]. Two dedicated fibers are provided to each household. One fiber handles the telephone (Plain Old Telephone Service, POTS) in a double star architecture, and the other supports a multiplex of video and ISDN services, with bidirectional facilities, in a single star architecture. (Note: in the final version of the system, these three types of services will be implemented on one fiber only, with a digital video switch at the hub.) The Optical Network Interface, at the subscribers' premises, and the video selectors handling 64 channels at the Central Office are custom designed and will authorize all foreseeable video services, but the amount of active devices needed makes this approach unlikely to be affordable in the near term.

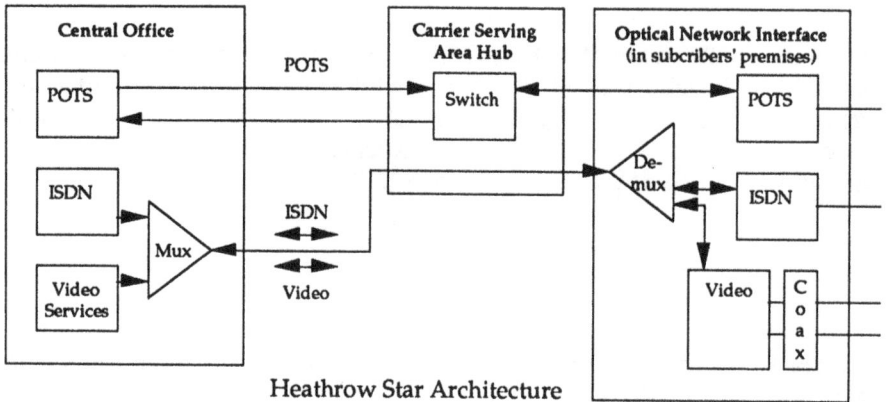

Heathrow Star Architecture

(After: Balmes, M., op. cit.)

<u>GTE in Cerritos</u>

GTE and Apollo Cablevision chose to implement two unidirectional fibers per household in a CATV system trial in Cerritos, California [26]. At the control center, 16 broadcast and four analog-switched video channels are digitized and combined with one ISDN channel and one control channel. Upstream, a 107 Mb/s bandwidth includes one video channel using a VCR or camcorder as a source, one ISDN channel, and one control channel. The position of the switching facilities at the control center introduces some delay to the subscriber. To help deal with this, information on the programs is made available as videotext, so as to reduce the amount of "zapping". This approach uses available technology and makes possible the implementation of many video services. However, it necessitates two light sources and two dedicated fibers per subscriber, all the way from the control center, the costs of which may still prove too high for actual use.

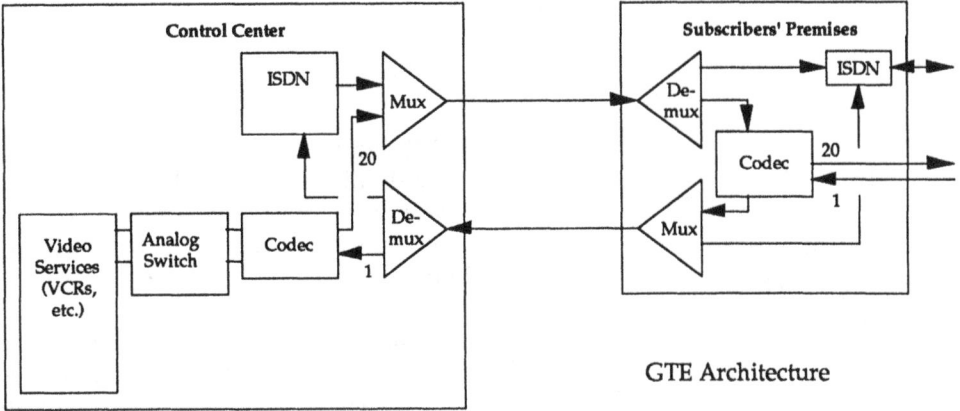

GTE Architecture

(After: Walker, S., op. cit.)

Bell Communications Research

One approach being proposed by BellCore [27] uses multiple wavelengths (colors) of the light to share the fiber among 20 subscribers. The fiber is taken to the home

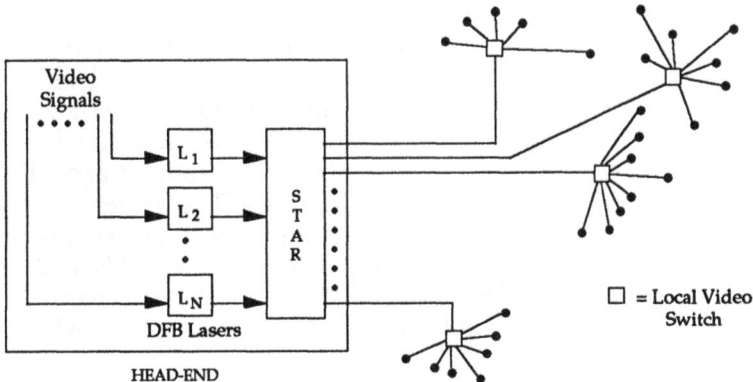

Bellcore Star Approach

(Source: Wagner, S., op. cit.)

in a star configuration using splitters to share the fiber cost. The published work provides for full bidirectionality of service and is therefore difficult to compare with other more conventional CATV schemes. However, with active optical components in each home it is bound to be somewhat expensive [28]. The system is not yet commercialized to our knowledge.

British Telecom

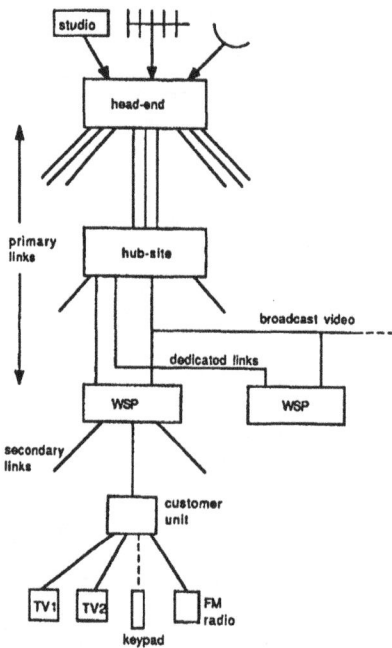

Switched Star Cable TV System

(Source: Fox, J.R., op. cit.)

British Telecom has been actively pursuing fiber optic distribution architectures since the early 1980s. Their work is now focused on two separate approaches which are both in field trial stages.

The switched star cable TV system [29] as illustrated in the figure, uses a distributed active "wideband switch point" to allocate channels to subscribers. Each subscriber has a dedicated coaxial link to the switch point and has the possibility of 2 simultaneous television channels and the FM radio band. Each switch point supports up to 300 subscribers and is contained in a 4-foot high outside plant cabinet.

The second approach is termed the passive optical network or PON. [30] This approach was pursued to try to overcome the high costs of the switched star system. It uses passive optical splitters and a time division multiplexing (TDM) technique [31]. The system is currently optimized for digital telephony. The optical signal is split 128 times to provide ISDN service to 128 subscribers [32]. British Telecom has built an experimental system to provide for CATV transport on the system to provide 32 channels.

Both of these proposed architectures are being tried in the same location in 1990 for about 480 subscribers in order to test technical and economic feasibility [33].

Raynet Corporation

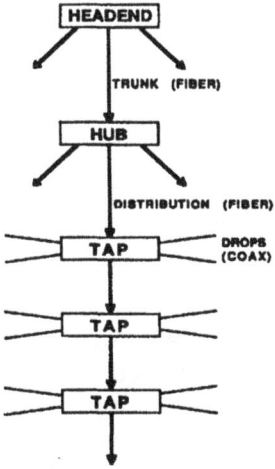

Raynet Corporation [34] has announced a fiber to the "curb" architecture that uses passive splitters to share the optical component cost over a number of subscribers. The cable usage per subscriber is estimated to be 70 feet, much the same as for the coaxial tree and branch architecture.

Raynet Star-bus Network

(Source: Large, D., op. cit.)

3.6 Hometown Cablesystem Architecture

At this point we play a trick on the reader. In the Hometown analysis we actually took a first pass at technology selection for the trunk system before looking at architectures. The choice was made because we already had an AM microwave trunk. If the system were all coaxial to start with, we probably would have started with an architecture analysis. So, if you want to understand how we ended up with an FM fiber trunk system to connect the head-end to our hub, you will have to go to Chapter 4. Since the process of planning is iterative between technology selection and architecture we can start with either.

After the first pass at the technology in Hometown Cablesystem we ended up with an FM trunk system linking the head-end to the hub location. We chose not to further optimize our technology selection at that point. The hub existed before for the AM microwave and we chose to leave it where it was. Our objective for the trunk was to get a SNR of 60 dB at the hub point. With the FM system we chose we are actually at 67 dB. We have some room to move in subsequent iterations if we can save any money by altering our choice.

Our current architecture is that we have a single non-redundant trunk and coaxial tree and branch distribution. We will take an architectural look at each of them using the strategies and methodologies outlined in Chapters 1 and 2.

Distribution

In Chapter 2 we split our budget over a hypothetical feeder and distribution portion of the system. We will now apply those budgets to see what type of architecture we will have.

A summary of the budgets is:

	Feeder	Distribution
Investment Objective	$16.90	$4.76
to meet: Signal Quality	55 dB	50 dB
Reliability (Outage Time)	0.36 hrs	1.42 hrs
Futures Investment	$2.98	$0.84

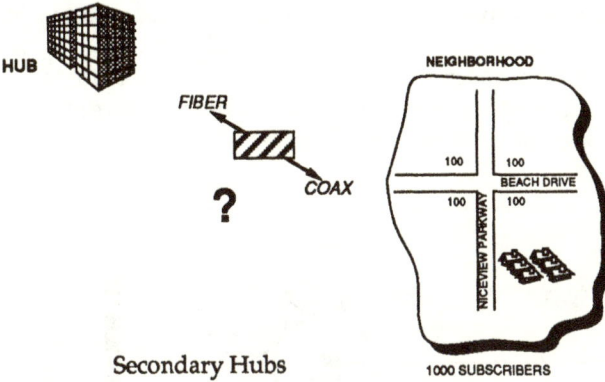

HUB

FIBER

COAX

?

Secondary Hubs

NEIGHBORHOOD

100 100

BEACH DRIVE

100 100

NICEVIEW PARKWAY

1000 SUBSCRIBERS

We first wanted to see how close we could get the fiber to the home given our budget. We did this by hypothesizing fiber placement to secondary hubs of 9000, 7000, 4700, 1000, 500, 125 and 6 subscribers. That is, for our system if it had 6, 8, 12, et.c, secondary hubs. To get the base price, we used the least expensive fiber technology option available, an AM system with either 12 or 18 channels per fiber, depending on the distance needed, and maintaining 55 dB at the secondary hub point.

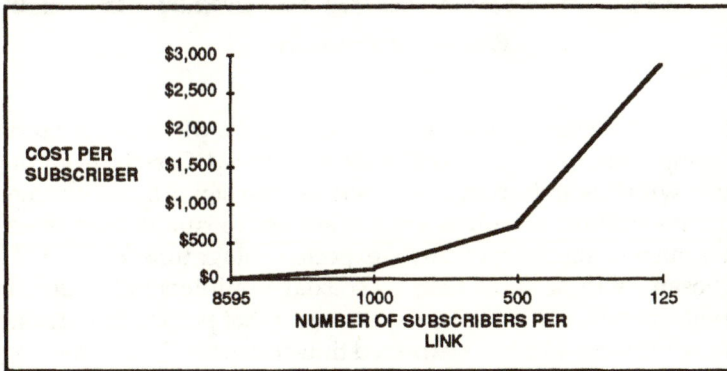

Pushing the Fiber Towards the Home

We found that for hubs of 9000 homes, we could implement fiber optics for $14 per subscriber, which fell within our $16.90 budget. The cost rises rapidly as we try to push the fiber closer to the home. To go to the 7000 home level, the cost would have been $19.88 which would have used up the entire feeder budget and left us nothing additional for futures. (Of course if we want to consider going to 7000 homes to be a better futures investment than 9000, we may do that.)

An analysis was also done to show the effect of using optical signal-splitting techniques at the hub point. The dominant cost in all cases was the electronics on the end of the fiber, not the fiber itself, so we thought we could reduce that dependency by splitting the signals to reduce the number of transmitters. This helps to some degree, but in our particular example, the distances were such that we could not afford the power used by these devices; i.e. the AM fiber system we chose did not have enough optical budget left to fully make use of them (See Chapter 4 for a discussion on optical link budgeting.)

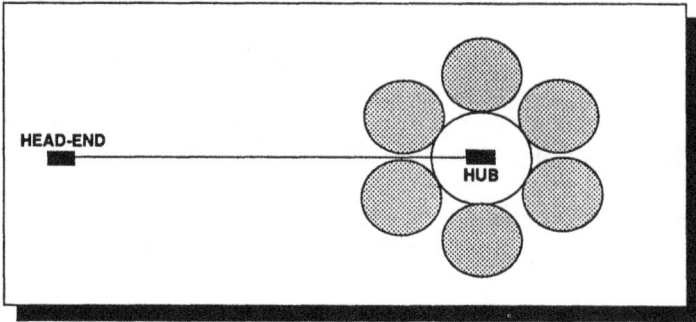

First Pass Architecture

Our first cut at architecture yielded six secondary hubs with 9,000 subscribers each. (The remaining subscribers to make 70,000 are served directly from the primary hub.) If we go with this architecture, we also meet our signal quality budget because our maximum amplifier cascade is actually six. For reliability we suspect we meet the objective but will need to confirm the expected outage time for the AM fiber links we have chosen. With no more than six coaxial amplifiers in the distribution we should be within the 1.42 hours of outage time for that portion of the system (using the 2% failure rate and a 10-hour expected time to repair). The remaining issue for our architecture at this point is the futures budget, which we will apply to fiber count.

Trunking

There are two things we can look at in the trunk portion of the plant: reliability and futures. By virtue of our selection of an FM trunk system, we already know we meet the signal quality budget and we are within our overall investment budget. From Chapter 2, we expect the following for the trunk system:

	Trunk
Investment Objective	$63.35
to meet:	
Signal Quality	60 dB
Reliability (Outage Time)	0.22 hrs
Futures Investment	$11.18

With the FM system only using up $13 per subscriber at this point, the budget for the trunk looks to be overly generous. We may wish to recycle our budget and apply more budget to the feeder or distribution if we need it, or reduce the budget somewhat.

One area to check is reliability. With a single unprotected link we are probably not currently meeting the 0.22 hour per year outage objective (99.997% availability) [35]. We could move to a higher level of reliability in two ways: adding automatic protection or providing a back-up route.

In Hometown, the cost of a back-up route came to about $15 per subscriber, considering only the extra cable needed (45 km) and some rudimentary optical switches at each end to effect the re-routing. Our money probably would be better spent on something to protect the laser as it is more likely to fail than the cable. At this point in the Hometown analysis we should leave architecture and go into a second round in technology selection to see if we can make any changes based on the architecture we have obtained.

Fiber Count for Hometown Cablesystem

<u>Trunk Cable Sizing</u>

The characteristics of the link are:

length	20 km
number of channels	60
technology	FM
number of channels/fiber	16

• Basic number of fibers: Four fibers at 16 channels per fiber will be more than enough for our 60 channels, no channel growth being expected in the next five years. Two spare fibers added, the cable size requires six.

• Reliability and survivability: The link is not protected, and does not support any secondary route.

• Future planning: We set a future deployment benchmark of $11.18 per subscriber for the trunk. In our 70,000 subscriber system, this means over $782,000 to be divided between future electronics placement and augmentation of the cable size. Fiber cost accounts for an estimated 20% of the total. Applied to our budget for the future on the trunk, we obtain $156,400, or $2.23 per subscriber, for extra fibers. An estimate of $1.00 per meter and per fiber (installed) allows us to add eight fibers on 20 km.

• Shared cable: No portion of the trunk is here shared with another link.

• Possibility of leasing: None was identified in the strategy.

We are now able to fill up our fiber count form (see figure). The optimal cable size for Hometown Cablesystem trunk is calculated to be 14 fibers. (Note: Since this does not correspond to a common cable modularity, and so we would likely change it to a 16 fiber cable in the end.)

Feeder Cable Sizing

We now take as a second example one of the feeders, with the following character-istics:

length	7 km
number of channels	60
technology	AM
number of channels/fiber	18

The signal is split optically after 4 km to service another secondary hub of very similar characteristics.

• Basic number of fibers: four fibers at 18 channels per fiber, and two spare make a basic size of six.

• Reliability and survivability: No addition is planned here.

	LINK DESIGNATION	
STRATEGIC FIBER COUNT FORM	**Tx End:** Head-End **Rx End:** Hub1 **Number:** Trunk 1	

Basic number of fibers		6
Reliability and survivability factor	+	0
Future	+	8
Foreign fibers sharing the cable	+	0
Optically shared fibers	−	0
Leasing	+	0
Cable size		**14 fibers**

Fiber Count Form: Head-End to Hub

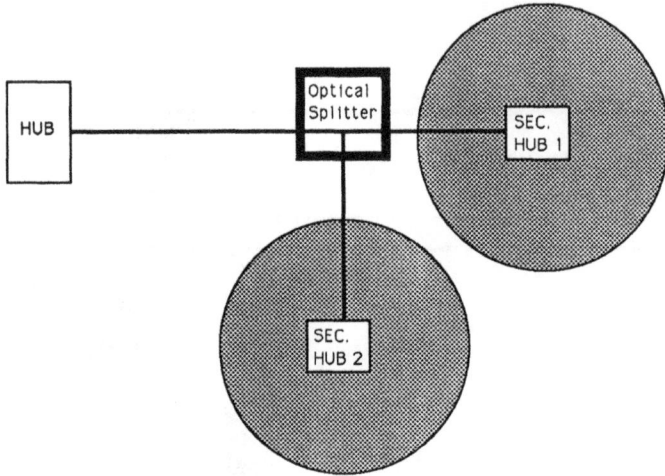

Optical Splitting of a Feeder

• Future planning: In our budget for Hometown, the future deployment bench-mark was set to $2.98 per subscriber for the feeders and hubs, meaning here $208,600. The total distance hub to subhubs being about 64 km we will need to spend $256,000 for the minimum four fibers assuming $1 per meter installed cost. This adds $3.66 per subscriber.

• Shared cable: No fiber will be subtracted, but since this is the first feeder sizing for the system, we will take into account the contribution of the other line: the two fibers we just added as a provision for the future are here physically part of the cable we want to size, upstream from the splitter.

• Possibility of leasing: None was identified.

The optimal cable size for Hometown cablesystem feeder to hub 1 is calculated to be 12 fibers. This is shown on the fiber count form.

STRATEGIC FIBER COUNT FORM	LINK DESIGNATION
	Tx End: Hub
	Rx End: SubHub1
	Number: 1 - 1st part

Basic number of fibers	6
Reliability and survivability factor	+ 0
Future	+ 4
Foreign fibers sharing the cable	+ 2
Optically shared fibers	- 0
Leasing	+ 0
Cable size	**12 fibers**

Fiber Count Form: Hub to SubHub 1

References

1. Baldwin, T.F., and D.S. McVoy, *Cable Communication*, Prentice-Hall,1988, p. 34

2. Lopez, J.A., Nynex Plans Test of Linking Homes with Fiber Optics, *Wall Street Journal*, February 23, 1989.

3. Robinson, D., Telcos in Cable - Overbuild or Acquisition?, *CED*, Special Supplement, November 1988, p. 22.

4. Baldwin, T.F., op. cit.

5. *The Costs of Telephone Company Installation of Fiber to the Home*, study by the Research and Policy Analysis Department of the National Cable Television Association, June 1989.

6. Quoted in *Fiber Optics News*, July 24, 1989.

7. While this expression has been used many times in the past to describe fiber optic bus systems, the most recent credit for it goes to Raynet Corp.

8. Garbanati, L.F., Bell Communications Research, Fiber Optics to the Subscriber: Objectives, Costs and Technology Assessment, *IEEE Proceedings of the International Symposium on Subscriber Loops and Services*, September 11-16, 1988, 14.3.1, p. 286.

9. Large, D., Raynet Corp., The Star-Bus Network: Fiber Optics to the Home, *CED*, January 1989.

10. Announcements by Aster Corp., quoted in *Fiber Optics News*, October 9, 1989.

11. Estimated.

12. Private source.

13. *Reliability Prediction Procedure for Electronic Equipment*, Technical Advisory TA 000-23620-84-01, Bell Communications Research, 1984. (Note: There may be a more current version.)

14. Cable placement is covered in depth in Chapter 5.

15. Some loss effects (0.02 dB per km) can start to effect single mode fibers after 25 years. In general, fiber cable failures are dominated by the "back-hoe blackouts". A good review article on the subject is: Szentesi, O.I., Reliability of Optical Fibers, Cables and Splices, in *Advances in Fiber Optics Communications*, H.F.Taylor, ed., Artech House, 1988, p. 57.

16. Canadian figures for aerial installation averaged from different providers. The renting cost of the poles is not taken into account in the installation, since in most cases the cable operators already have accounted the right of way in their coaxial cable network cost. As for the price of the optical cable itself, for example purposes in the course of the guide we will use 40 cents per fiber plus $2 for the sheath, per meter.

17. In general, systems available today for CATV are not protected.

18. This summary is from notes taken during talks by W. Ciciora and J. Chiddix of ATC given at the Western Showcase in Anaheim, December 7-9, 1988. An overview article of the approach is: Chiddix, J., Fiber Backbone: A Proposal for an Evolutionary CATV Network Architecture, National Cable Television Association Technical Papers, 1988, *Cable '88*, p. 73.

19. Luff, R.A., The Broward Cable Area Network Fiber Model, *CED*, February 1989, p.27.

20. Rogers Cable TV of Toronto has issued a number of documents relating to their proposed fiber optic plans. This summary is from *Project 94 - Cable's New Horizons*, Canadian Cable Television Association, February 1989, p. 25.

21. per discussions with G.Bradley of SaskTel in Regina, Saskatchewan.

22. Kaiser, P., Fiber-to-the-Home, *Optics News*, October 1989.

23. Alcatel Network Systems, Raleigh,NC, private correspondence, December 1988.

24. Robinson, D., op. cit.

25. Balmes, M., The Technology behind Heathrow, *Telesis*, 1989, vol. 16 no. 2, p.30.

26. Walker, S. (GTE Laboratories), paper presented to the Electronic Industries Association, Advanced Television Committee, November 6, 1989, Washington, D.C.

27. Wagner, S., WDM Applications in Broadband Telecommunications Networks, *IEEE Communications Magazine*, March 1989, p. 22.

28. Garbanati, L.F., op. cit.

29. Fox, J.R., *Evolution of the British Telecom Switched-Star Cable TV System*, National Cable Television Association Technical Papers, 1989, *Cable '89*, p.119.

30. Oakley, K., Passive Fibre Local Loop for Telephony with Broadband Upgrade, *British Telecommunications Engineering*, Vol.7, January 1989, p. 237.

31. For the uninitiated, time division multiplexing is a telecommunications technique which

allows for multiple subscribers to simultaneously use the same link without interfering with each other. This is done by allocating each subscriber a "time slot" within a signal. A master clock determines who uses the facility and when. As long as the clock works fast enough, everyone can use the facility without anyone else knowing (up to some reasonable number).

32. ISDN stands for Integrated Services Digital Network, the main attributes of which from the transmission point of view seem to be putting 144,000 b/s on a 12,000-foot copper pair. While a challenge for copper, this is somewhat trivial in the fiber world leading some wags to proclaim that ISDN really stands for "I Sure Don't No", or "Innovations Subscribers Don't Need". At any rate ISDN has considerable benefits in other aspects of telephony than transmission.

33. Rowbotham, T., (British Telecom Research Laboratories), *Proposals for Local Loop Optical Field Trials*, February 1989.

34. Large, D., op. cit.

35. The manufacturer of the FM system in question estimated the availability to be 99.993% which certainly is not that far off. If the feeder or distribution portion of the system turn out to be more "available" than we think, we may be able to accept this and save the cost of the alternate route or protection scheme.

Chapter 4
System Parameters and Technologies

4.1 Introduction to Fiber Optics: Strategic Issues in Optical Systems

The key areas affecting television transmission in fiber optics are:

- Noise and linearity problems, related to analog-signal modulation;
- Bandwidth limitations in digital systems;
- Passive devices related issues.

Noise and Linearity in Analog Systems

When an analog format is chosen for signal transmission, the signal quality at the end of the link is very dependent on the linearity of the source, and on the amount of noise generated in any point of the link.

Digital Channel Capacity

Format conversion is a main challenge in digital transmission. A bandwidth of 90 Mb/s is necessary in order to transmit the full information contained in an NTSC television signal. Down from this figure, multiple compression techniques are used, and one channel now can be contained in as small a bandwidth as 45 Mb/s, corresponding to the DS3 rate used in the telephone network.

Here, also, each manufacturer develops its own solution, and no standard compression algorithm has yet been chosen by the standard bodies.

However, in classical digital transmission, the cost of the equipment is prohibitive, compared to analog modulation.

Issues Related to the Use of Passive Components

The possibility of using passive components to modify easily a topology, or save money on active components, is restrained by the amount of optical power required by these operations, i.e., in most cases, the signal is attenuated to the point where it can no longer be transmitted with a decent quality.

This will become more apparent with the presentation of the different components involved in optical transmission systems and their mode of operation.

4.2   Introduction to Fiber Optics: Components of Optical Transmission Systems

Whatever the chosen system, from any manufacturer, a transmission link based on optics will consist of the same basic elements, as illustrated in the figure, "Typical Optical Transmission System". Functionally, an electrical signal stimulates a device which emits light, the source; the light travels down a waveguide, the optical fiber, and is received by a detector which creates a current reproducing the input signal. A certain number of passive devices are needed between the source and the detector to connect fibers, or to split and couple the light.

The optical link elements are described here only to the level sufficient for optical transmission characterization. Optical cables and mechanical properties of the optical medium are treated in Chapter 5.

Please note that the performances included here are accurate to the best of our knowledge, but so much effort is concentrated today on transmission optics enhancement that, from the very beginning, we could announce all figures as being obsolete.

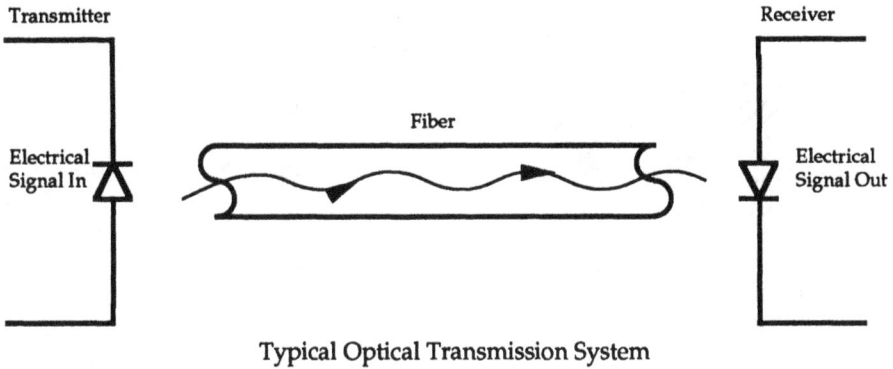

Typical Optical Transmission System

Optical Fiber

The optical fiber is a strand of solid glass used to conduct light for communications purposes. In the fabrication process, the fiber is drawn from a one meter glass rod to lengths of several kilometers and ultimately has the diameter of a human hair.

Optical fibers are currently used on transmission distances that range from a few centimeters (such as their projected use in a car's electrical network) to hundreds of kilometers (transcontinental and transoceanic installations [1]).

The optical "waveguide" is created by varying the refractive index in the manufacture of the optical fiber. The refractive index is a measure of the relative speed of light in a vacuum (i.e. free space) to the speed of light in a particular medium (here optical fiber) [2].

The basic optical fiber consists of two concentric layers of glass (see the figure titled "Cross Section of a Fiber"):

• The core, the inner layer, serves as the medium for light transmission. The core has a higher refractive index than the surrounding cladding.

• The cladding provides a lower refractive index in order to contain the light in the core to facilitate transmission down the fiber: the light injected into the core strikes the outer edge of the cladding at angles defined by the difference between the refractive indices, and is reflected back into the core.

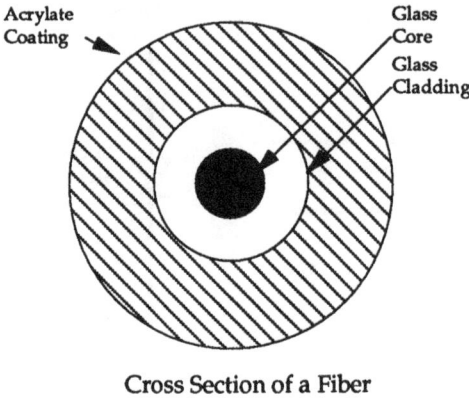

Cross Section of a Fiber

The quality of the glass used in optical fibers reflects impressive technological achievements, and causes entertaining comparisons such as:

• if the sea were made of optical fiber glass, we could see clearly down to the bottom of the deepest fault in the Atlantic Ocean,

• or, let us say, what you see through sun glasses is approximately equivalent to a 3-m thick window, and also to a 30-km thick piece of optical fiber glass, etc. [3].

Parameters of Transmission in Fibers

Optical fibers can be described for transmission through the following parameters:

• Attenuation

Attenuation, or fiber loss, is the decrease of signal power with distance in the fiber, expressed in dB/km.

The range of loss depends on the configuration of the fiber: material, size core/cladding, optimization for a frequency, etc. Further loss can be induced by microbending and microirregularities in the waveguide.

In the range of spans we are considering, attenuation is proportional to fiber length. It does not increase with modulation frequency, but depends on the wavelength in use [4]. In communication applications fiber optics operate in the near infrared spectrum, dimensioned as illustrated on the figure, "Scale of Wavelengths for the Range of Known Electromagnetic Waves".

The fibers' peculiar profile of attenuation led to the choice of three basic transmission wavelengths over fiber, also referred to as "fiber windows":

10^{-4}	◄— Gamma rays
1	◄— X-rays
	◄— Ultraviolet
	◄— Visible
10^4	◄— Near Infrared
	◄— Infrared
10^8	◄— Short radio waves
10^{12}	◄— Radio broadcasting
(Wavelength in nm)	◄— Wireless

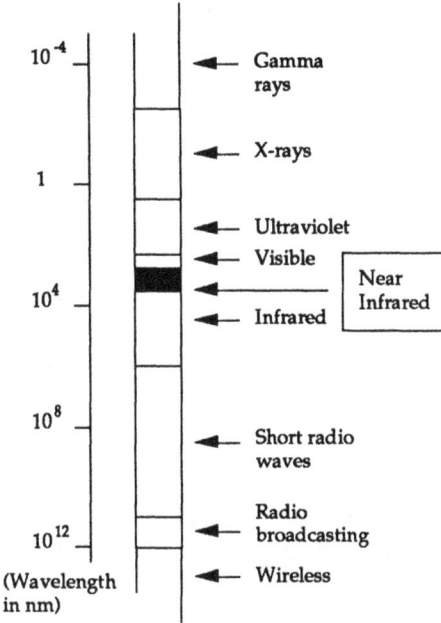

Scale of Wavelengths for the Range of
Known Electromagnetic Waves

(Source: Jenkins, F., *Fundamentals of
Optics*, Fourth Edition, McGraw-Hill,
New York, 1976. Reprinted with
permission)

- 780-850 nm, provide low loss, and are known as "short-wavelengths"; they deal with "first generation" optics,

- 1310 nm, provides even lower loss and low dispersion,

- 1550 nm, provides the lowest loss and dispersion.

These last two are known as "long-wavelengths", and deal with "second generation" optics.

Optical fibers are engineered so they present zero-dispersion for a specific window: i.e. they can be "dispersion shifted" for a precise wavelength. High quality "double-window" fibers are dispersion nulled for both 1300 and 1550 nm.

The research driven on light sources was focused first on the short wavelengths, they being easier to engineer. Today, highly efficient sources and receivers are available in the long wavelengths as well.

Note: in the line of our comparisons to the benefit of optical fiber glass quality, we could say here that an impurity of transitional metals can induce an attenuation of 2000 dB/km when its concentration is one part per million (1 ppm), i.e. one atom of impurity for 10^6 atoms of glass; that concentration is of the same order as the concentration of the impurities in a very pure semiconductor (intrinsic) [5]. This is to be compared to the 0.3 dB/km currently achieved.

• Modal Dispersion

When the core diameter is not small enough, several waves of light are propagated down the fiber, effectively following different paths, of different lengths. This gives rise to a spreading of light over time as it travels through the fiber.

• Chromatic Dispersion

If the light injected in the fiber contains more than one wavelength another dispersion mechanism takes place: signal spreading from different velocities of different wavelengths.

The solution to chromatic dispersion is found in very good quality light sources, with narrow emission spectrum [6] (also called narrow linewidth sources).

• Bandwidth

The spreading of the light along the fiber through dispersion effects eventually leads to the loss of the information. The bandwidth of a fiber can be defined by the highest frequency signal which may be effectively transmitted, and thus is a measure of the information-carrying capacity of the fiber.

It is measured at the half-power point (-3 dB). For fiber specification, it is referenced to one km, and expressed in MHz-km. It is usually given only at a particular wavelength.

For example, a fiber declared to have a bandwidth of 600 MHz-km can carry up to 600 MHz of data on one kilometer, or 300 MHz of data on two kilometers, etc.

• Numerical Aperture (NA)

The numerical aperture of a fiber defines the maximum angle at which a fiber will accept light and propagate it [7], as illustrated on the figure "Acceptance Angle and Numerical Aperture".

High NA means ease of coupling of the light into the fiber, but also greater dispersion: as NA increases, the bandwidth of the fiber decreases. A good match of NA of system components minimizes loss in a fiber optic system.

For a high efficiency, joining low NA fibers with very precise sources are best.

• Intrinsic Fiber Loss

Intrinsic fiber loss is the loss contributed by variations in the manufacturing of the fibers:
> Core-diameter variation,
> Core-to-cladding eccentricity,

Core concentricity,
NA variation.

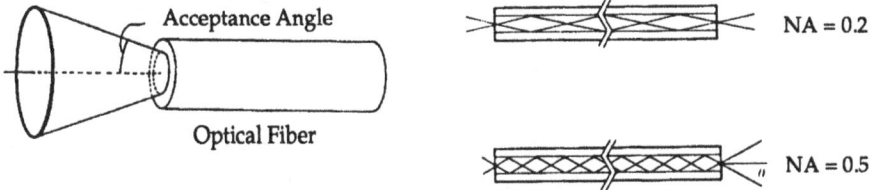

Acceptance Angle and Numerical Aperture

(Source: Nérou, J.P., op. cit. Reprinted with permission.)

Types of Fibers

Fibers are usually classified by their refractive index profiles and their core/cladding sizes. There are three main types of fiber:

- Multimode Step Index Fibers,
- Multimode Graded Index Fibers,
- Single-Mode Fibers.

These three types are illustrated in the figure "Physical Characteristics of the Three Main Types of Fibers".

• Multimode Step Index Fibers

All multimode fibers have a large core diameter, allowing several waves of light to propagate through the fiber.

Multimode fibers were the first ones to be manufactured; their advantage is the low cost of all system components, but their characteristics make them useful only on relatively short distances. Today, they are most commonly employed in local area network (LNA) applications.
Step index fibers are the easiest to engineer; they induce though the highest dispersion characteristics of all fiber types.

Core Profile	Light Propagation	Refractive Index Profile

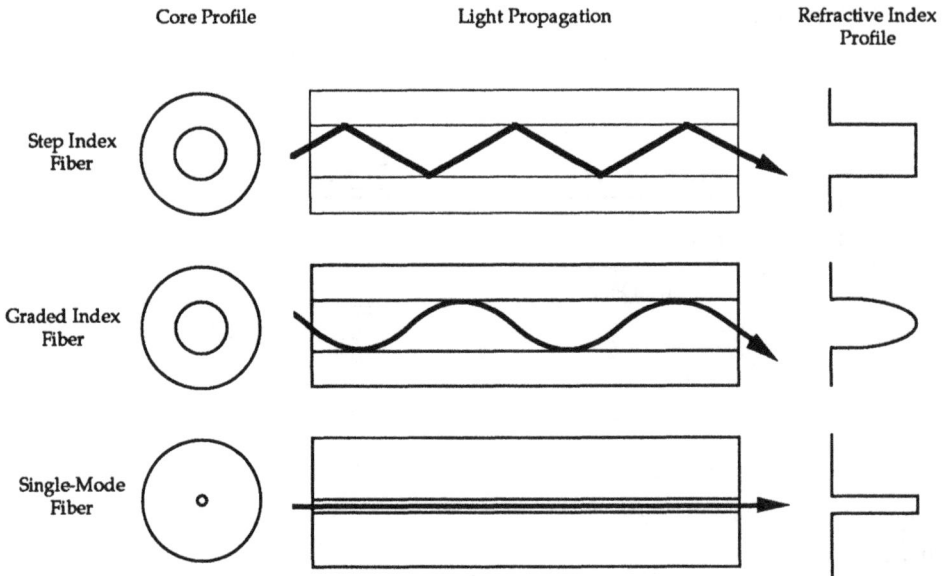

Physical Characteristics of the Three Main Types of Fibers

Step index fibers are further classified by their material composition:

- All glass,
- Plastic-clad silica (plastic cladding and glass core),
- All plastic.

• <u>Multimode Graded-Index Fiber</u>

After step-index fiber, graded-index fibers were designed to improve the dispersion characteristic: the core is here a series of layers, each with a successively lower index of refraction, continually bending the light towards the core axis.

Dispersion is reduced, and the bandwidth much broadened, but they are delicate to manufacture.

• Single-Mode Fibers

Single-mode fibers are step-index fibers with a core diameter small enough to allow only one light mode, above a specific wavelength ("cutoff wavelength"), to propagate.

Dispersion is reduced to a few picoseconds per km. This allows such a huge bandwidth that the fiber itself is commonly said to be 20 years in advance over optical electronics.

The drawback lies in the precision required for all manipulations: coupling of the light into so small a medium (asking for laser sources), connecting, splicing; the technical progress in this domain has been very fast in the last years, and the problem now stands more in the selection of the equipment.

All this make single-mode transmission the most adequate solution to CATV types of problems.

Type of Fiber	Core Diameter (μm)	Cladding Diameter (μm)	Numerical Aperture	Nominal Attenuation (dB/km)	Bandwidth (MHz-km)	Frequently Encountered Sizes (core/clad., in μm)
Multimode Step-Index						
All Glass	50 to 400	125 to 500	0.15 to 0.4	< 50	< 25	100/140, 200/250...
Plastic-clad silica	50 to 500	125 to 800	0.2 to 0.4	< 50	< 25	-
Plastic	200 to 600	400 to 1000	about 0.5	< 1000 at 650 nm	-	-
Graded Index	30 to 75	100 to 250	0.2 to 0.3	< 10	< 200	50/125, 62.5/125...
Single-mode	3 to 10	50 to 125	about 0.1	< 3 at 850 nm < 1 at 1500 nm	< 2000	9/125

Table of Typical Fiber Characteristics

(After: *1988/89 Buying Guide*, Fiberoptic Products News, p.T54)

Passive Devices

A passive device is an optical component which does not require active powering for operation; i.e. connectors, splitters, couplers, and the fibers themselves. These generally have excellent temperature and lifetime characteristics but introduce some loss into the system.

These losses have been very much improved, and a lot of work is still put on this issue to minimize them in the direction of transmission (forward loss) and maximize them in the opposite direction (return loss). Though, high frequencies and bidirectionality have put an added burden: reflection of the light back into the laser.

Two different types of connection can be performed in an optical link: for a removeable connection, a connector will be added; for a permanent connection, a splice will be done.

At a point of connection, the total loss includes:
        NA mismatches
        Diameter mismatches
        Connector or splice insertion loss

Splices

Three main types of splices exist in order to connect two fibers in a definitive way:

• Fusion Splices

Fusion splices are the best in terms of
            robustness (70% of the original waveguide strength),
            return loss (60 dB achievable),
            forward loss (around 0.1 dB),
            reflections [8].

Their drawbacks are found in
            cost,
            difficulty of realization on the field: It is a specialized craft, requiring sophisticated fusion and test equipment; a solution can be found in contracting services.

• Mechanical Splices

Technical characteristics depend very much on the executant, for example, for rotary splices, the return loss is found between -20 and -57 dB depending on the person.

Forward losses are comparable to fusion, however, and it is a more flexible splicing method, requiring less expensive equipment and training, and is less restricted by environmental problems. The fine tuning is done through the test equipment.

• Vee-groove Splices, without Fusion

In vee-groove splices, two waveguides are fitted into a vee-groove and butted.
Forward loss is here higher than for fusion, 0.2 to 0.3 dB.
Very fragile, these splices have to be roughened with strength members.

The technology stay within these three types, but is evolving rapidly. The characteristics are to be checked regularly.

Choice is here left to the user, and is dependent of the application: choose mechanical splices for ease of installation when loss and reflections are not too much of an issue. Choose fusion splices when they are more critical.

Connectors

Connectors are used to link the fiber to any device, when splicing is not desired. They are removable, thus adding flexibility to a link.

Many types have been developed; most of them are based on butt-coupling of the cleaved fiber ends which allows direct transmission of optical power from one core to the other. The alignment has to be as precise as in the case of a permanent connection, but the problem is worsened by the disconnectibility requirement, especially with the single-mode, small-cored fibers.

Some 70 different types exist at the moment, and losses range from 0.1 dB to 1.0 dB depending on connector type.

Reflection was an important defect in connectors a short time ago; new families now offer much lower reflection figures, down to - 30 dB to - 40 dB (the PC or physical contact families). Mechanical characteristics of connectors will be described in more detail in Chapter 5.

Standard names are applied to short cables sold with connectors already in place:

- Jumper cables, or patch cords, have identical connector plugs on both ends of a short cable.

- Pigtail cables have a connector plug on one end of a cable, the other end is bare fiber.

- Adapter cables have two dissimilar connector plugs, one on each end of a cable.

One type of connector should be chosen according to the links requirements, and be the only one used for the engineering in order to simplify the implementation and reduce the number of adaptations.

Couplers-Splitters

Sometimes referred to as a "combiner", the same device can combine or split the light from one to two or more ports. A coupler is a bidirectional device, and half of it is a true image of the other half: a 1xN coupler will have the same characteristics as a Nx1 one.

Star couplers with multiple input and output are available in multimode up to 64x64, and are wavelength independent. But single-mode couplers are phase sensitive because of the single wavelength, and it is very difficult to fabricate good couplers larger than 2x2.

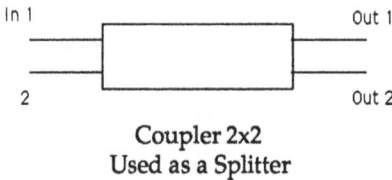

```
In 1 ─────────┌──────────┐──────── Out 1
              │          │
    2 ────────└──────────┘──────── Out 2
           Coupler 2x2
         Used as a Splitter
```

Standard single-mode couplers are usually characterized for one wavelength because performances are wavelength dependent. However, so called wavelength independent couplers can now operate on a larger spectrum as for example from 1300 nm to 1600 nm [9].

The couplers come with a one-meter fiber, usually pigtailed, on each port (the fiber type is thus to be specified).

On top of low forward loss and high reflection attenuation, a low level of crosstalk between the ports is expected from a coupler. The main parameters are:

• Coupling Ratio

It is the percentage of output signal transferred through each port. The coupling ratios can be specified from 1 to 99%. For example, common shares are 50%/50%, 70/30, 90/10...

• Directivity

The directivity quantifies the amount of reflection and of crosstalk generated inside the coupler. This parameter is also named near-end isolation.
> 40 dB for multimode
> 50 dB for single-mode

In a coupler, all unused ports must be properly terminated, or will generate reflections in the system.

• Insertion Loss

The insertion loss is the total loss introduced by a coupler, from a link point of view, i.e. it is measured from the input port to the output port we are using. The insertion loss includes the effect of signal splitting or coupling, and is not only a connection loss.

For example, for a standard 2x2 single-mode coupler used as a splitter, insertion loss can be estimated in the following manner:

50 / 50 split loss	3 dB
Extra loss in splitter ("Excess" loss)	0.25 dB
Provision for nonuniformity	0.25 dB
Total	3.5 dB

As an indication, on a 64x64 multimode coupler, insertion loss has been announced to 21.5 dB [10].

The price discrepancy on couplers performing the same functions is high, depending on the excess loss performance, tolerance on the splitting ratio, and stability in polarization [11].

However, if using a splitter can save a transmitter, because the light beam from one transmitter will be used on two links, even the highest priced one would easily prove in. Yet, the insertion loss makes this type of system somewhat difficult to engineer.

When possible in terms of loss, the use of couplers can bring a lot of functionalities in a link:

- reduce fiber, source, connector, or splice source,
- interconnect nodes,
- mix signals through wavelength division multiplexing:

• WDM

WDM stands for wavelength division multiplexing. It allows the transmission of different signals, modulated on different wavelengths, on the same fiber, thus increasing the capacity of fiber links and adding to network flexibility. Today, WDM is ususally done in order to upgrade systems operating at 1300 nm with a second channel at 1550 nm.

The multiplex of two wavelengths can be realized with simple couplers; specialized WDM devices can handle two or more, up to ten wavelengths [12].

The insertion loss of this type of coupler can be as low as 0.5 dB.

Optical Isolators

When the reflection in a transmission link is too high (we will see in the discussion of light sources, for example, that this can be an important factor), it can be useful to add an optical isolator. The following are an example of an isolator performance, at 1300 or 1550 nm [13]:

Fiber to fiber insertion loss: typ. <= 1.5dB, max. <= 2.0 dB
Isolation: >= 35 dB
Return loss: >= 60 dB

Active Devices: Light Sources

The price of the equipment at the transmit end is usually much higher than that at the receiving site [14]. This comes partly from the difference of complexity and price between light sources and detectors. Both are active devices, meaning that they require active powering for operation.

Qualities of Light Sources

The research of the optimal light sources is driven by the physical limitations introduced by the optical fiber; we can define the basic ideal qualities of a light source to be:

• Smallness: Have a small light emitting surface, smaller, if possible, than the core of the fiber, to maximize the light transmission,

• Directivity: Emit the light on directions compatible with the acceptance cone of the fiber. The directivity of a light source is illustrated by the "Radiance Distribution Pattern", which plots the light power in function of the emission angle,

• Appropriateness in terms of wavelength: Emit on wavelengths corresponding to the minimum attenuations of the fiber,

• Precision in wavelength: Minimize the chromatic dispersion thanks to a small spectral width.

Other qualities include:

• High power output,
• Short response time,
• Ease of modular packaging,
• Reliability (independence from temperature, etc.),
• Small cost.

Numerous devices are available for converting electronic signals to light; however at present only two classes of device are suitable for fiber optic systems:

• Light-emitting diode (LED)
• Injection-laser diode (ILD)

Both are based on fiber optic components using the electroluminescent [15] characteristics of man-made semiconductor crystals such as:

- GaAs (gallium arsenide),
- GaAlAs (gallium aluminum arsenide),
- InGaAsP (indium gallium arsenide phosphide).

Both lasers and LEDs have their unique advantages. The final choice between them in any application depends on the characteristics of the system.

Light Emitting Diode

In the light-emitting diode family, we can distinguish

- Surface-emitting LEDs,
- Edge-emitting LEDs.

Their common characteristics are the following:

- Strengths:
    Ease of manufacture,
    Low cost,
    Long lifetime,
    Reliability (not too sensitive to temperature variations).

- Weaknesses:
    Bandwidth limited to several 100 MHz,
    Low optical efficiency: the input current is translated into an optical output which is much lower than the theoretical maximum,
    Not monochromatic.

This last characteristic means that LEDs emit several wavelengths, and it is their most important weakness: their spectral width is between 25 and 60 nm.

When LEDs are used as sources for analog transmission, large signal excursions cause nonlinear operation because of the normal nonlinearity of the diode characteristic curves. Efficient techniques, involving feedback circuits, have been developed for this purpose.

- Surface-emitting LED's Characteristics

Surface-emitting LEDs are mostly:
    based on GaAs,
    of limited power, from 100 W to 2 mW,
    presenting a source radiance distribution pattern very flat.

This is a second weakness of SE-LEDs: nondirectivity, as seen in the figure: the light is spread in every direction.

Thus, the main use of SE-LEDs is found in multimode links, involved, for example, in local area networks.

• Edge-emitting LED's Characteristics

Edge-emitting LEDs are:
> • often based on GaAlAs,
> • of higher emitted power, from 0.2 mW to 8 mW,
> • more directive, as illustrated.

EE-LEDs are structured like lasers, but cannot attain the level of drive current necessary to experience the laser effect.

They are useful for short-range single-mode systems, because their narrow beam angle can achieve a good coupling with the fiber. Their reasonable cost could eventually make them suitable for fiber-to-the-home applications.

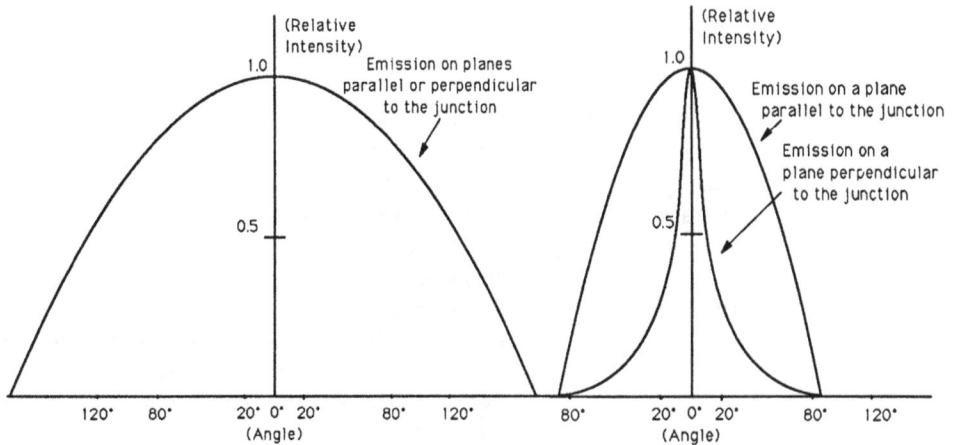

SE-LEDs (left) and EE-LEDs (right)
Radiance Distribution Patterns

(After: Nérou, J.P., op. cit.)

Injection Laser Diode

Lasers are the most suitable sources for long-distance single-mode transmission and they are most commonly used for CATV applications.

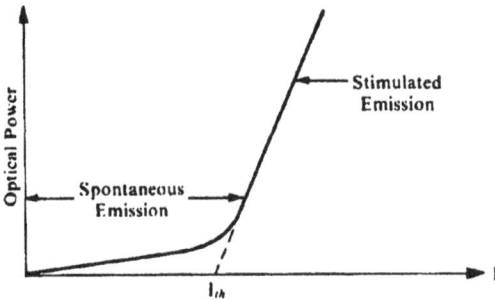

The Laser Effect

(Source: Baker, D.G., op. cit. Reprinted with permission.)

Laser stands for "light amplification by stimulated emission of radiation". The laser effect occurs above a certain value of the input current, the threshold current, and is represented by a sudden change in the slope of the emitted power as a function of drive current. Threshold current value vary from 15 to 100 mA, depending on the laser structure.

The emission wavelength depends on the semiconductor chosen. Two important categories are:

GaAlAs/GaAs --   emitting around 820 nm,
GaInAsP/InP --   emitting between 1100 and 1600 nm.

A range of characteristics include:

Bandwidth --   from 400 MHz to 10 GHz,
Optical efficiency --   much higher than LEDs',
Optical power --   from 1mW to 40 mW,
Spectral width --   very narrow (see the figure, "Typical Emission Spectra of Injection Lasers and LEDs"): 2 or 3 nm, 0.1 nm recently,
Source radiance distribution pattern --
very directional. Lasers couple the light much better in single-mode fibers, as shown on the figure, "Laser Diodes: Radiance Distribution Pattern".

These characteristics are improved over the ones displayed by LEDs. In counterpart, drawbacks are:

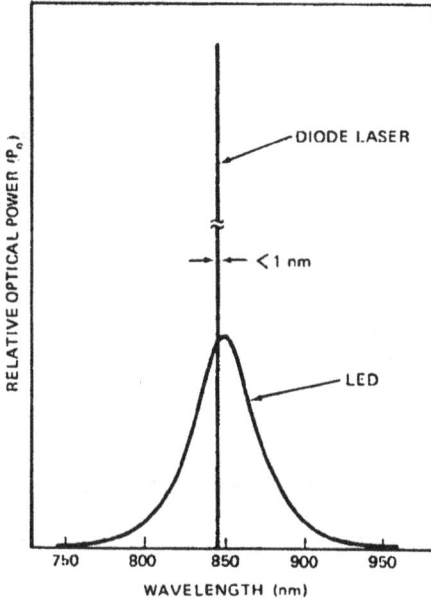

Typical Emission Spectra of Injection
Lasers and LEDs

(Source: *Electronic Design Magazine,* Vol. 28, No. 8,
Penton Publishing, 1980. Reprinted with permission.)

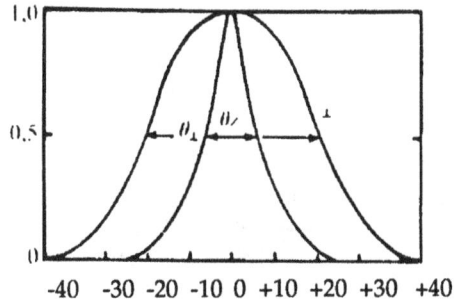

Laser Diodes
Radiance Distribution Pattern

(Source: Nérou, J.P., op. cit.
Reprinted with permission.)

• Price [16]

• Sensitivity to Optical Reflections

Random reflections occur in every element of the optical link, to a certain degree:
back into the laser cavity they can cause laser instability (intensity and wavelength
fluctuations). These fluctuations introduce noise in the system. Minimizing reflections is thus a big issue in optical elements construction.

For better coupling and less reflection, the transmitter usually comes already
connected to external fibers: a "pigtail" from the external connector presents a fiber
for splicing or low loss connection to fibers. Typically, coupling loss between
emitter and fiber range from 0.1 to 1.5 dB [17].

Reflection problems can be solved when an optical isolator is used directly after the laser, whenever the link loss budget can admit it.

• Sensitivity to Temperature

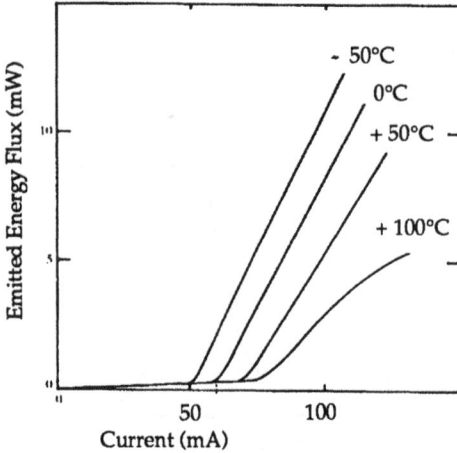

Current (mA)

The immediate effect of temperature is an increase in the threshold current and diminution of the slope efficiency: i.e., the laser becomes less accurate and the power consumption increases. The characteristic called tracking error is a measure of the influence of temperature in dB.

Effect of Temperature on a Laser

(Source: Nérou, J.P., op. cit. Reprinted with permission.)

• Shorter Lifetime

The lifetime of a laser is a function of temperature and level of operation.

Deciding upon this tradeoff is a key issue in laser operation.
Aging has the same type of influence on the characteristic curve of a laser diode as temperature has. Upon manufacturer's advice, the lifetime can be calculated on the basis of an increase in threshold current of 50%, or, some threshold currents being low (15 mA), on the basis of an absolute increase of it (+30 mA, for example) [18].

Thus, on any laser module, the output must be monitored and feedback control is provided in order to maintain the output power constant and the drive current variation in the most linear region, mainly through:

• Automatic Gain Control (AGC)
• Temperature Control Electronics.

These control circuits make the system more complex and therefore less reliable; their effect is limited to a range of values (e.g. 10 dB for AGC [19]).

A laser today can expect a lifetime of between 20 and 50 years [20].

A few more words on two trendy lasers:

- Distributed FeedBack Laser (DFB)
- Fabry-Perot Laser (FP)

Both are injection lasers, the noise characteristic of which has been improved for a better signal quality in transmission, with some difference in the way the light is amplified inside the laser before the emission to the outside.

• DFB Laser

It is now a favorite in AM communication for cable TV because of improved linear characteristics, giving way to a bigger bandwidth: more channels can be transmitted over a single fiber.

Recently DFB lasers with a spectral width of only 0.1 nm have become available. Other improvements are a low threshold current (20-50 mA) and a very low internal noise. Very sensitive to back reflections, their packaging can include optical isolators, protecting the laser and allowing some improvement in coupling efficiency.

Not all the DFB production is usable for AM performance requirements. Output power is currently up to 5 mW, and a lot of work is being put on it to improve it: it would then be able to serve more than one fiber at a time, reducing a cost which is currently its defect.

• Fabry-Perot Laser

It delivers new and improved characteristics too, and is less expensive than the DFB laser [21], but is far more limited in terms of bandwidth [22]. Also, its second order performance is slightly poorer than the DFB laser and can be a limiting design factor.

In conclusion, better linearity, narrow emission spectrum, and higher available power are the main qualities that make laser diodes preferable over LEDs for most CATV applications.

Active Devices: Optical Detectors

The Perfect Optical Detector would be:

- Sensitive,
- Quick,
- Small,
- Of low cost,
- Insensitive to environment (temperature, etc.),
- Of low noise.

Semiconductors are used one more time, in detectors named photodiodes, which create an electrical current in response to light input.

- GaAs and Si (Silicon) are used around 800 nm,
- Ge (Germanium) and its composites, InGaAsP, are used for longer wavelengths.

Transmission characteristics of detectors include:

- <u>Responsivity of a Photodiode</u>

It is defined as the ratio of a detector's output current versus the input light power, in Amperes per Watt. For example, a responsitvity of 0.5 A/W means that by receiving an optical energy of 1 mW, the photodiode will respond with a 0.5 mA output current.

- <u>Sensitivity of a Receiver</u>

This important notion for link design is the minimal input optical flow, expressed in dBm, needed to achieve a given signal quality; it depends on the responsivity of the photodiode and the noise. From this minimal flow we can derive the maximal length of a link. The minimum sensitivity ranges today from -30dBm to -50dBm typically [23].

- <u>Internally Generated Noise in a Photodiode: Optical Noise</u>

So called shot noise comes from the random variation of the output current of the photodiode around its average value. Dark noise comes from the saturation current of the photodiode, depending on the type of diode (for example, germanium is 10,000 times noisier than silicon [24]) and on the temperature. It can be measured when no light is received, and the detector is in the dark. Both can be minimized, but will always be present to some degree.

Three main types of devices are suitable for use as detectors:

- PIN diodes
- Avalanche photo diodes (APD)
- PIN field-effect transistor (PIN FET)

Each diode is optimized for a given wavelength.

PIN Diodes

The most simple and most used light detector is a PIN diode, where PIN means that between doped parts P and N lies an intrinsic (nondoped) zone. Responsivity of a PIN diode varies between 0.5 and 0.8 A/W.

The PIN diode is inexpensive, easy to use, and its bandwidth is sufficient for a range of applications, though it does not provide any internal amplification of the signal. It is always associated with a preamplifier [25].

Avalanche Photo Diodes

APDs have inherent gain associated with the detection mechanism through an electron multiplication process or "avalanche".

This avalanche process increases the amount of noise generated internally , and the multiplication factor of the internal gain is very sensitive to temperature, but responsivity of the diode is improved by a factor between 10 and 100.

They also can respond faster than traditional photodiodes, and thus allow transmission at higher frequencies.

APDs improvements over PIN diode are responsivity, and bandwidth (up to

1GHz).

Drawbacks include: increased power requirements, greater cost (around 10 times that of PIN diodes), fast saturation in received power, need for compensation circuits for protection, and higher generated noise.

However, for many applications, electrical noise is much greater than optical noise; in these cases, an APD will much improve the signal quality when implemented.

An optimal multiplication factor can be found for each application: an APD amplifies both the incoming signal and the incoming noise; this improves the signal quality as long as the dominating noise is the electric noise generated by the surrounding electronics. The optimal is reached when amplified incoming noise and internal noise become comparable to the electric noise.

PIN FET

The PIN FET also has an internal gain. It presents though some linearity defects in its characteristic that prevents it from being used for analog modulated signals. Even in digital systems, its lack of speed usually limits it to low bit rates [26].

Other Active Devices

External Modulators

Instead of a modulation of current on the source in order to produce a variable light beam, another way of transmitting a signal is to modulate the light itself in phase, amplitude or possibly frequency. A range of "external modulators" exist, mainly as components developed on LiNbO3 (lithiumniobate, a cristalline ceramic which features low loss and high electro-optic capabilities and is widely used today in electro-optic components).

The principle of the process is simple: a waveguide is created in the lithiumniobate substrate. When applying an electrical field across the waveguide a refractive index change occurs, resulting in a phase modulation.

When two waveguides exist, and the signal is recombined after the phase modulation, the interferometric effect creates an intensity modulation as, for example, in a Mach-Zehnder modulator [27].

External modulators still know some distortion problems and their insertion loss is high (around 3 to 5 dB), but both research labs and manufacturers are working on this promising approach.

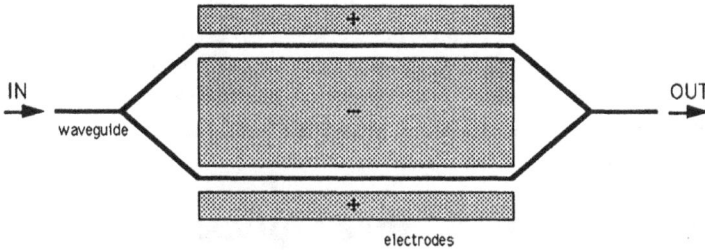

Example of a Mach-Zehnder Modulator

Optical Switches

Future wideband communication systems may require direct optical switching of the signal, without conversion back to electrical. Besides manual switches, three main types can be identified [28] : electromechanical switches, waveguide switches (made by electric, magnetic or ultrasonic signals), and switches made from a secondary material. All are independent of frequency and code format ("frequency and code transparency") [29].

Electromechanical switching involves a movement: rotation of a prism, lenses, mirrors, or of a fiber, with the use of electrical energy. Insertion losses were typically 1.5 to 3 dB, but are getting much lower [30]. However, the movement is relatively slow (around 10 ms) and the simple fact of it being mechanical could possibly be a weakness for long term operation.

The other two types operate at much higher speed, and are useful as well for integrated optics applications, which, after some more improvements, seem the only possibility in high-speed transmission. Integrated circuits for optical switching are built mainly on LiNbO3. The maximum size of an optical switching matrix is limited by fabrication techniques (up to 10x10) and their use by the tolerable insertion loss and the total crosstalk [31]. None are in production yet that we know of.

Optical Amplifiers

Optical amplifiers can be used as on-line amplifiers or preamplifiers for receivers [32]. Typically, they provide a gain on the signal of 15 dB. They are modulation and frequency transparent, multichannel, and have a very large bandwidth, as advantages over regenerators, but introduce some noise and dispersion in the signal, and are polarization dependent. Polarization adjusters and associated feedback circuits may be added to overcome this dependency. Nevertheless, some recent works promise a gain of 30 dB and an improvement in polarization independence that would permit them to do without polarization adjusters for a number of applications [33]. A lot of progress can be expected in this field.

4.3   Introduction to Fiber Optics: Transmission System Considerations

Important parameters which define the level of performance of a fiber optic link are the signal quality, addressed by the carrier-to-noise ratio (CNR) or the bit error rate (BER), and the distortion parameters (second-order products, third-order products, intermodulation, etc.).

Signal Transmission with a Laser Diode through Direct Modulation

An operational laser ideally features low noise, high speed, and high linearity.

In the most common mode of optical communication, the direct modulation of the laser, the drive current is modulated in order to transmit a signal. To keep the operation in the linear portion of the characteristic curve of the laser, the mean value of the current is increased to a biased value, ranging from around 15 to 80 mA. Two important characteristics for a laser under operation are:

• Optical Modulation Index (OMI),
• Relative Intensity Noise (RIN).

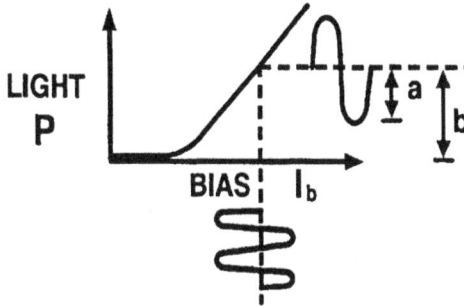

Modulation of a Laser

(Source: Way, W., *Subcarrier Multiplexed Lightwave Systems*,
Optical Fiber Communication Conference, 1989 Tutorial
Sessions (Optical Society of America, Washington, DC
1989), p.55. Reprinted with permission.)

• OMI is equal to 1, or we speak about a 100% depth of intensity modulation of the light when an RF signal that is superimposed to the bias reaches the lasing threshold with its peak (on the graphic, OMI = a/b). The modulation depth is then the relative size of the current driving the laser (i.e. the bias) versus that of the modulated signal. For the same power, OMI is proportional to the square root of the number of channels in the input signal [34].

A bigger OMI means a stronger signal, but there is a tradeoff between OMI and nonlinear distortion: for lasers that are directly modulated, the modulation is usually 50% or less. Although some lasers presently allow for a modulation index of 60%, overmodulation for a few cycles can possibly result in laser source destruction [35].

• RIN is the noise specification of the laser. It can be considered to be the noise floor, measurable with a spectrum analyzer after light to current conversion with an optical detector.

RIN varies with the bias current value and thus with the emitted light power. It is also highly dependent on the amount of light that is reflected back into the laser (eventually, an optical isolator may prove useful). A good laser diode can have a - 145 dB/Hz RIN number. DFB lasers can have RIN as low as -155 dB/Hz.

For transmission with a laser, RIN is an important component in the calculation of

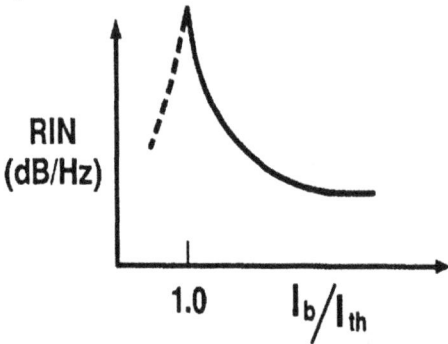

RIN
(dB/Hz)

1.0 I_b/I_{th}

Relative Intensity Noise in
Function of Bias Current

(Source: Way, W., op. cit.
Reprinted with permission.)

the Signal Quality: the lower the RIN, the better the CNR.

As an indication, in the not very practical case where OMI = 100% and 1 channel only is transmitted, the RIN gives directly the peak CNR, in a 1 Hz bandwidth [36]: CNR= -RIN - 3, in dB.

Optical Performance Parameters

Optical performance parameters are here

- Signal Quality:
    Carrier-to-Noise Ratio,
    Signal-to-Noise Ratio,
    Bit Error Rate.
- Nonlinear Distortions:
    Second-order products level, and Composite Second order (CSO),
    Third-order products level, and Composite Triple Beat (CTB).

Signal Quality

Different parameters are used according to the transmission format:

- BER

In digital modulation, the signal quality is measured as the bit error rate, which defines the error ratio on the recognition of the coded signal at the reception; the most common value taken as a reference is 10^{-9}. There is a relation between the BER

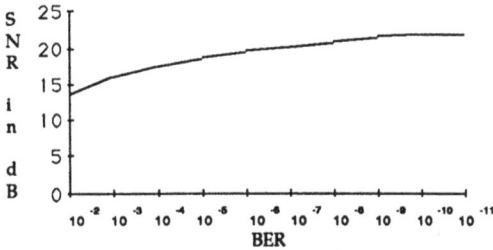

Relation between BER and SNR

(From: Baker, D.G., op. cit.)

and the Signal to Noise Ratio as it is used in analog modulations for signal quality measurement [37].

The comparison between signal quality in digital modulation and in analog modulation is still a matter of discussion for the standards bodies. The heart of the problem is the quantization error, which is inherent to the format: it is the difference between the real continuous values of the signal and the discrete values it is encoded into. The fact of accounting this error as noise or leaving it out of the calculations makes a noticeable difference on signal quality measurement (from 54 dB to 60 dB, for example).

• CNR and SNR

In analog modulations, the parameters used to measure the signal quality are the carrier-to-noise ratio and signal-to-noise ratio.

The CNR is a function of [38]:

- modulation depth per channel,
- laser noise,
- quantum noise,
- absolute temperature,
- receiver noise.

- Contribution of the modulation depth:

As more channels are added to the system and are transmitted through the same source, there is a share of modulation, and the modulation depth diminishes.

- Contribution of the noise and temperature:

The contribution of the different factors vary according to the length of the line (i.e. the amount of loss between source and detector) and the type of modulation chosen.

An FM optical transmission system can be limited by the laser, the quantum noise, or the receiver:

Laser limited:

In a single-mode system, the CNR contribution of the laser transmitter is constant, function of the Relative Intensity Noise of the laser, and does not vary with distance. On *short optical links* (typically 10 to 25 km, this is the main case encountered in CATV transmission) the CNR performance of the receiver is so much greater than the CNR performance of the laser transmitter, that overall link CNR performance will be dominated entirely by the CNR performance of the transmitter. The RIN term predominates in the equation, making a laser with low RIN essential, and the biggest problem is the laser sensitivity to reflections. The use of a PIN detector is here adequate.

Quantum noise limited:

It can be the case in *medium length optical links* (typically above 25 km). If the detector has to be an APD, because of amplification needs, it would be wise to compare the noise contribution of different semiconductor APDs: Ge APDs are more noisy than InGaAs ones, for example.

Receiver limited:

The CNR contribution of the optical receiver is directly related to the input optical power. Hence, receiver performance is proportional to distance and the CNR performance of the optical receiver will decrease until it is below that of the transmitter. The receiver noise is then the limiting factor in *long length links* (typically above 30 km).

The figure, "Contributions to CNR in FM Systems", shows the two main domains of the CNR curve as a function of distance.

SNR Calculation in FM Systems

When CNR is known, we can derive SNR:

SNR link = SNR optics + SNR electronics   (the + indicates "power" addition)

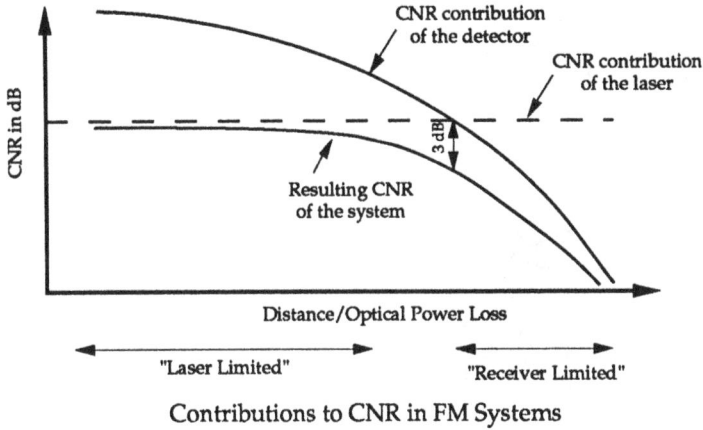

Contributions to CNR in FM Systems

(After: Holobinko, J., American Lightwave Systems, in Seminar
ADCOM Electronics Ltd., Toronto, March 7, 1989.)

SNR electronics:

The SNR of the FM electronics vary by manufacturer. It is equal to the baseband SNR performance of the given electronics connected electrically back to back, without any fiber link in between.

A minimum number can usually be provided.

SNR optics:

In FM, SNR optics = CNR optics + FM processing gain (pre-emphasis and postemphasis)

The FM processing gain, also referred to as FM enhancement factor is added by the mere utilization of FM modulation on the signal. It is proportional to the deviation used in the modulation. Each manufacturer can specify it for his own equipment [39].

SNR in AM Transmission

AM transmission links are not laser limited. AM being very sensitive to noise, most of the time it is adequate to opt for a PIN detector.

In AM modulation, the two expressions of noise performance, CNR and SNR, are almost the same, about:

$$SNR = CNR - 0.5 \ dB$$

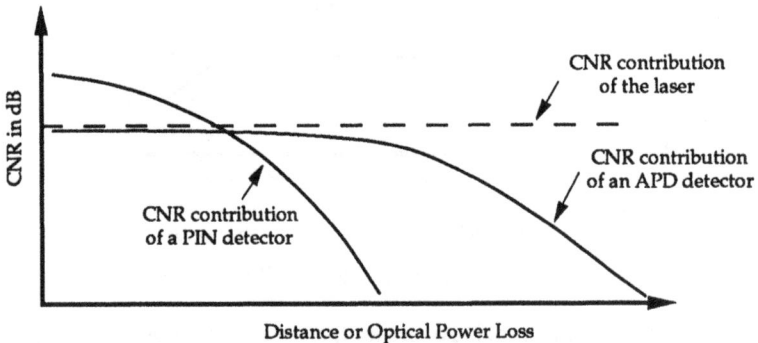

Contributions to CNR in AM Systems

(After: Holobinko, J., American Lightwave Systems, in Seminar
ADCOM Electronics Ltd., Toronto, March 7, 1989.)

• <u>Variation of the Point of Operation of a System</u>

The point of operation of an optical system is a choice between video signal transmission quality, number of transmitted channels, and optical loss budget.

Our system planning scheme is based on the arbitrary choice of video signal quality, among other performance parameters; signal quality being fixed, it is useful to be able to approximate number of transmitted channels in function of loss budget:

Some manufacturers give away simple equations, or curves that can be used as guidelines [40]. For example, the figure, "Point of Operation of an AM Optical System", represents the loss budget in function of the number of channels, at 51 dB CNR, in an AM modulated system.

Nonlinear Distortion

The nonlinearities of the laser transfer characteristic produce distortion in the signal in an analog system, when multiple channels are transmitted. The more channels

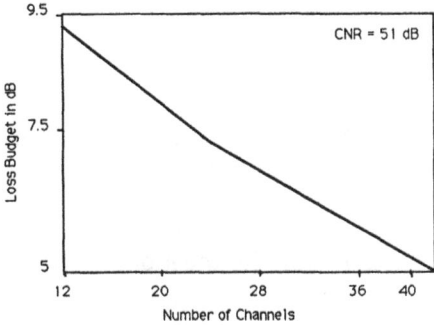

Point of Operation of an AM System

(After: Presentation Document for Laser Link™, Anixter Bros., Inc. Canada.)

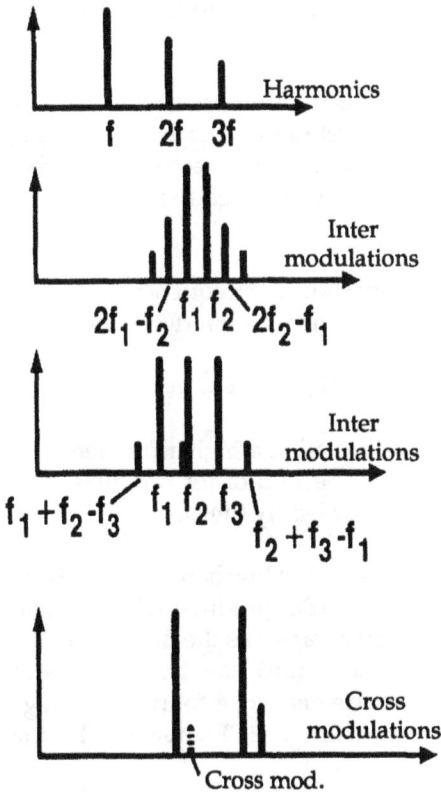

Nonlinear Distortions

(Source: Way, W., op. cit. Reprinted with permission.)

in the system, the more harmonics and intermodulation products are created.

Note: Some characteristics given by manufacturers will be measured with only two frequencies interacting: much less intermodulation is then taken into account than in full load.

• Second-order Products

Single-ended amplifiers once had second-order distortion that created the biggest limitation for CATV systems. The introduction of push-pull amplifiers solved that problem.

Lasers suffer from the same second-order limitations, and for the moment are not available with better characteristics.

Recommendations for the levels of discrete second-order interference are -60 dB by the NCTA [41].

As it happens with third-order distortion, the once-unimportant discrete second-order beat, when summed with many other discrete second-order beats falling on the same frequency (due to other channel pairs), results in a composite second-order distortion (CSO) which may have a level large enough to interfere with the visual carrier.

- In *vestigial sideband* (VSB) transmission, when standard frequencies are used, the second order sum products cause very visible luminance interference.

Three main approaches are used to minimize it [42]:

> The *octave system:* the channels are converted to a higher frequency where they do not exceed an octave anymore. Second-order products fall then above or below the frequency range occupied by the channels.

> The *split band system:* one laser can transmit the low band (54-88 MHz) and the high band (174-216 MHz). All second-order products fall above, below, or between the two bands. A second laser transmits the mid band (120-174 MHz) , which is less than an octave. And a third laser transmits the super band (216-294 MHz). If more channels are needed, a fourth fiber might be used for the hyper band, as presented in the figure, "Frequency Plan for AM Transmission".

> Second-order beats can then be easily filtered out by a RF passive filter placed at each receiver right after conversion from light to RF, but before combining with the other fiber signals. These bandpass filters would also reduce the noise bandwidth, useful property, four fiber pathes introducing

One fiber for 24 ch. | fiber 1 | fiber 1 |

Two fibers for 36 ch. | fiber 1 | fiber 2 | fiber 1 |

Three fibers for 54 ch. | fiber 1 | fiber 2 | fiber 3 | fiber 1 |

Four fibers for 72 ch. | fiber 1 | fiber 2 | fiber 3 | fiber 1 | fiber 4 |

Channels	2 6 7	24 25	36 37	54 55	72
Frequency (MHz)	54 108 120	228	336	408	516
	audio fm				

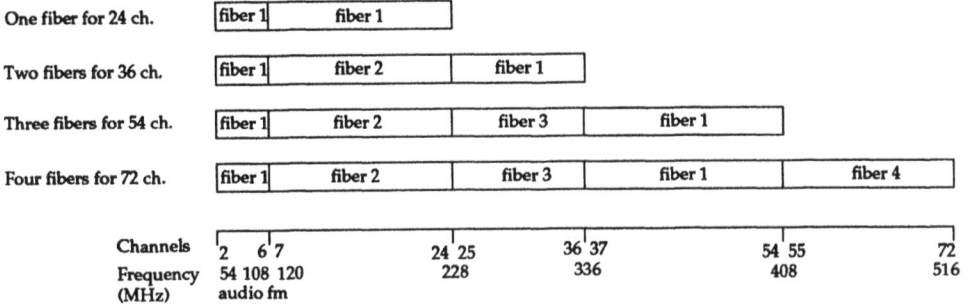

Frequency Plan for AM Transmission

(After: Lamarre, A., *Les Technologies Relatives à la Fibre Optique*, Presentation Document for Fiber Optics Related Products, Cabletel Communications Inc., 1989.)

more noise than a single one.

HRC: harmonically related carriers. All carriers are locked to a common phase reference, and second- and third-order products fall with no frequency offset on top of visual carriers. When phase noise is low the visibility of such an impairment is very low.

- In *wideband FM* transmission, the solution can be found in choosing carrier frequencies so that the second-order products fall in between the signals. The third-order products then become the predominant distortion, and are spread over the whole frequency range [43].

Each manufacturer develops its own solution, like:

12 channels, 100-540 MHz, channel spacing 40 MHz; the 2nd order harmonics are halfway between video carriers. Audio carriers are centered around 60 MHz, second-order harmonics are interleaved between video carriers [44]; or

16 channels, 60-510 MHz, channel spacing 30 MHz; the audio carrier is 9.5 MHz below the video carrier [45].

• Third-order Products

Third-order distortion is generally specified through the composite triple beat ratio (CBR). The limit of perceptibility for composite triple beat is considered to be around -60 dB.

Laser design is predominant in the effort to minimize third order distortion (this does not necessarily affect second order products).

Other Considerations

Interactions between Audio and Video

Transmission of stereo audio signals over fiber can be done using subcarriers or carriers at discrete frequencies. Discrete audio carriers are normally preferred because there is no interaction between audio and video [46].

With subcarriers the transmission scheme of audio has to be worked out to prevent these interactions, but in some systems video bandwidth can be reduced due to the subcarrier filters necessary to prevent crosstalk between video and subchannel signals [47].

In all cases, it is better to have guaranteed audio quality measured when both audio and video are on.

Mean Time Between Failure (MTBF) of Transmission Equipment

The surrounding electronics, control circuits and others, can substantially lower the lifetime of a module, simply because of their MTBF, in some cases, very low compared to the laser.

For example, using a laser quoted at 50-year MTBF, the whole module build around can drop to 25 years. Then the choice of the power supply is important: its MTBF can vary from 30,000 hr (rectified power supplies) to 100,000 hr (switching power supplies) [48]. Power backup is a good improvement factor.

External Modulation

External or indirect modulation techniques use a source that is continuous and either disrupt the beam (intensity modulation) or changes the wavelength (optical version of FM). This type of modulation transfers the linearity problems from lasers to external modulators, which are usually either electro-optic (interferometers) or acousto-optic. These devices still experience heavy second- and third-order distortions; third-order seems in a good way to be solved, second-order remains worrying, too much to claim the immediate availability of this technique.

External modulation sometimes uses specialized sources like high-power YAG lasers (which is a nonsemiconductor laser made of yttrium-aluminum-garnet). The fixed laser source operates at a much higher output power than in direct modulation (recently demonstrated up to 30 mW), thus lowering link costs: the beam can be split between links.

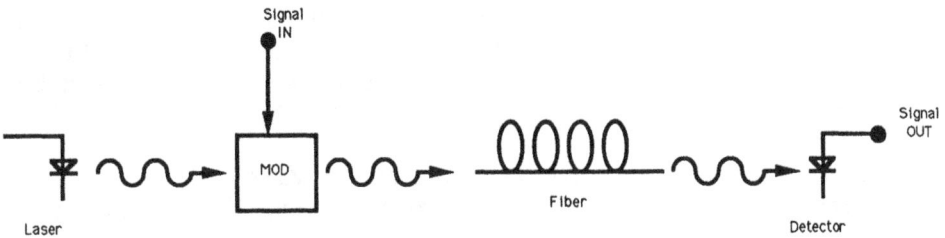

Principle of External Modulation

The external intensity modulation approach may inject some fresh blood in AM modulation, by using conventional sources combined with devices such as the Mach-Zender interferometer. The AM signal can be impressed on the light beam as an interference pattern. This is apparently Dylor's approach, although they are reluctant to release details [49]. Aiming at the set of requirements identified by ATC, the remaining problem seems to be the second-order products, with the following characteristics for the system:

- 60 channels/fiber,
- 54 dB SNR,
- 18 km.

• *Coherent modulation* means the modulation of a perfectly coherent source. It is currently studied by several labs. When light sources emitting only one wavelength will be available, the multiplex of multiple modulated carriers will allow the full use of the fiber optic bandwidth.

4.4   AM, FM, Digital Systems for CATV

Analog modulation (amplitude or frequency), and digital modulation systems are the choices available today. Their main characteristics will be described here.

Amplitude Modulation (AM)

Vestigial Sideband amplitude modulation is very attractive for CATV systems on fiber, because this is already the established modulation for television broadcast transmission. The existing network facilities would then easily connect to the new pieces of equipment, smoothing the transition from coaxial to fiber; this makes it the least expensive alternative.

Plus, AM is bandwidth efficient, requiring only 6 MHz per channel, and is transparent to all scrambling techniques currently used in TV transmission [50].

The counterpart is essentially its short reach and its sensitivity to noise and to all transmission impairments; the new SNR requirements are difficult to achieve, for AM modulation does not improve signal quality over CNR.

Substantial improvements made the AM approach more eligible:

• lasers with lower intrinsic noise (down to -155 dB) pump up the CNR: DFB [51] and Fabry-Perot lasers are now the basis for AM video transmission systems on fiber;

• elements with lower backreflection noise are developed to avoid the necessity of an optical isolator; still, all unused ports in optical elements (splitters) have to be properly finished;

• better knowledge of intermodulation mechanisms aid the development of (partial) solutions in form of frequency planning.

Today's AM achievement can be estimated as appears in the first column of the Table of Characteristics of AM Transmission (the typical price is quoted from [52]).

Another set of parameters has been issued as a condition point for the use of AM over fiber, and the implementation of fiber itself, by American Television & Communications Corporation, second column [53].

	Typical parameters	ATC requirements
Channels / fiber	40	60
Span length	10-12 km	18 km
SNR	51 dB	55 dB
CSO	60 dB	65 dB
CTB	65 dB	65 dB
Cross-modulation (XMOD)	60 dB	65 dB
Price	US$30,000	< US$5,000

Table of Characteristics of AM Transmission

Increase of the span length is here an important issue, because signal quality is delicate to achieve; and the repeating process (back to electrical, amplification and optical retransmission) somewhat costly in signal quality, would preferably be avoided.

The "typical" characteristics met as of today by amplitude modulation systems are pretty good relative to coaxial. For example, in our Hometown Cablesystem, we have one hub which we feed from the head-end using a fiber optic supertrunk to get at least 60 dB SNR at that point; from this 60 dB point, the 10 km typical AM system would yield around 51 dB SNR at its output. If instead, we had chosen to build the link with coaxial, the 51 dB SNR point would very likely have been reached much sooner; after only about 4 or 5 km.

Frequency Modulation

Compared to AM, FM appears much more costly. Each baseband video signal is FM modulated on a separate carrier frequency (audio and data services are either modulated on subcarriers and placed within the video signal, above 4.5 MHz [54, 55], or are modulated on separate FM carriers), meaning that each receive site has to perform the channel by channel demodulation back to AM for further distribution of the signal, thus needing a room with a civilized environment. This space requirement would prevent FM to the home now and for some time further.

Also, FM is not very bandwidth efficient: 30 MHz at least are required per channel. The "unlimited" bandwidth of single-mode fibers eases this part of the problem, allowing high-deviation FM and thus enhancement of SNR performance over CNR.

Like this, FM presents by nature much greater immunity to noise and transmission impairments. Overall signal quality is much easier to achieve, and on a much longer distance, providing that, as in AM, frequency planning is foreseen for low inter-modulation transmission.

FM encounters the same difficulties in repeaters than AM: some loss in signal quality because the linear characteristics of the repeater transmitter will be different from the original transmitter [56]. However, repeating is not here as important, because the FM enhancement factor of SNR over CNR already authorizes much longer spans for a given signal quality.

A delicate problem is that of transmitting scrambled signals. Scrambling systems that use baseband scrambling have to provide a clamp pulse that can be transmitted on a subcarrier to restore dc of the scrambled video signal at the receiver [57]. More difficult is the transmission of RF-scrambled signals.

As for today, it is best to test in lab the desired scrambling scheme on the system chosen prior to employing it [58].

Digital Transmission

Digital transmission is a type of "nonlinear modulation". The signal quality is not to be measured the same way as in analog modulation, and discussion is still open on the straight application of the standard EIA RS-250B, first designed for analog applications.

Assets of digital transmission appear to be

• uniform performance unaffected by transmission distance or system configuration,

• no signal degradation when repeated: If signal repeaters are required, decoding to baseband signals is not required; the digital data stream is simply regenerated after the receiver and transmitted into the second link,

• the dominance of digital technology in major telecommunications and signal processing industries which has created component and design advances. For example, telephone industry requirements are driving the development of systems at data rates of multiple gigabits per second,

• Low maintenance costs,

• Configuration flexibility (time division multiplexing for transparent mixing or drop-and-insert of video, audio and data services),

• Good foundation for future advanced television developments,

• Supply of integrated transmission of audio, video and data.

The main drawbacks are:

• Small video channel capacity per fiber, compared to analog; the bandwidth requirements are bigger than in FM. However, advances in coding schemes and optical technology (example WDM) may change the economics of this factor,

• Requirements in room and environment at receive sites, that are the same as FM: big,

• Peer incompatibility: DS3 codecs are currently incompatible with other manufacturer's DS3 codecs, because each uses a different compression algorithm,

• HDTV may not be compatible with today's 45 Mb/s NTSC Codecs [59],

• Incompatibility with scrambled signals [60],

• Price: last but not least, price of a typical digital system is much higher than that of analog ones; however, new approaches are reducing this difference [61].

4.5   Link Budgeting

What is a Link Budget?

All link parameters decided, topology outlined, homework done on active and passive components, this is where the engineer wants to check if the network of his dreams is really feasible. Optical link-loss budgeting systematically takes into account the loss every element of the link applicate on the transmitted optical power [62].

The optical power profile along a link can be expressed as in the graphic titled "Example of Optical Power Profile on a Link".

Elements of Link Loss Budgeting

The different elements of link-loss budgeting will be reviewed along with a budget example following the link described on that same graphic; the results are inscribed on an "Optical Link Budget Form" as illustrated.

• Launch Power and Receiver Sensitivity

Transmitters and receivers usually come with a connector output, or a pigtailed one, to minimize the losses between the fiber and the active elements. Thus, generally the manufacturers quote the launch and receive powers at the connector, or at the end of the pigtail.

The launch power is the optical power injected by the transmitter in the fiber. The receive power is also known as the receiver sensitivity; it is the minimum signal power needed to achieve a certain output quality (in digital, the BER, usually choosen equal to 10^{-9}, 10^{-11}, 10^{-12}; in analog modulations, the CNR).

The difference between the launch power (what is injected into the link) and the receiver sensitivity (what is needed at the end of it) is the end-to-end fiber budget, the maximum amount of power the link transmission may consume, and still be "safe".

In the example, the transmitter output power is -10 dBm, and the receiver minimum

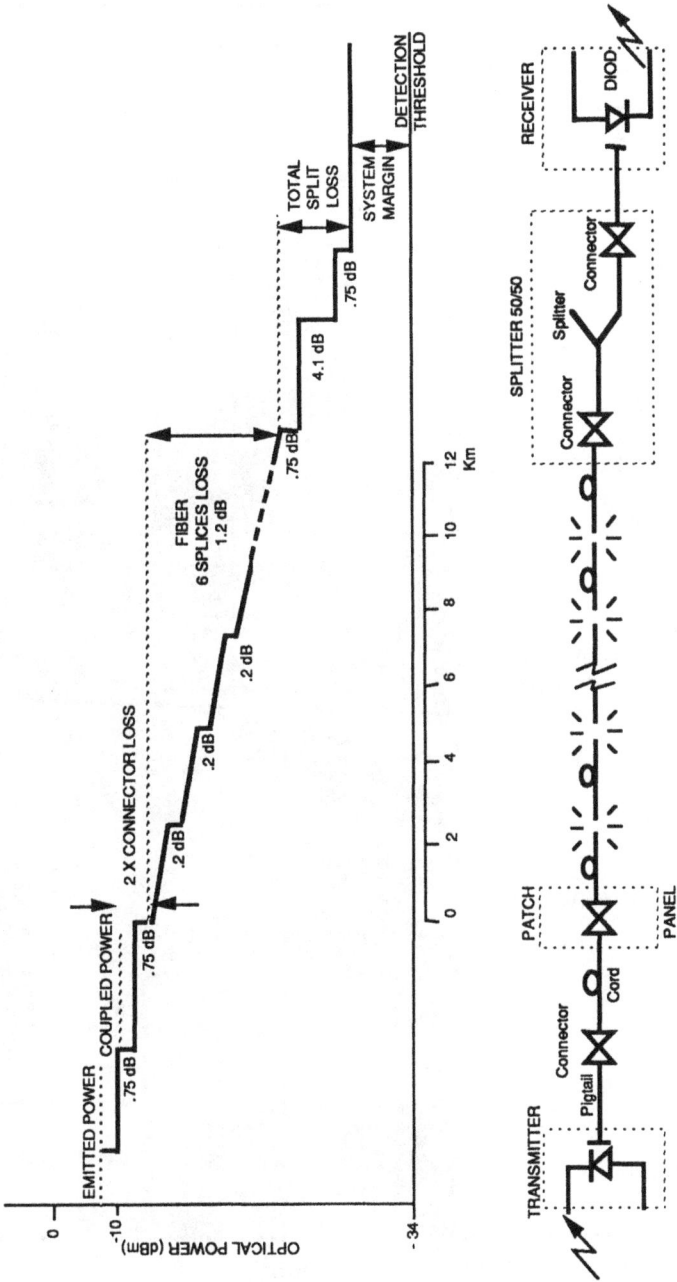

Example of Optical Power Profile on a Link

OPTICAL LINK BUDGET WORST CASE ANALYSIS	LINK DESIGNATION Tx End:_____ Rx End:_____ Number:_____	Nb of channels/fiber:_____ Nb of fiber links in the system:_____

1. END-TO-END FIBER BUDGET	ENGINEER'S CHOICES	MANUFACTURERS SPECS	POWER LEVELS
1.1 Guaranteed output power at transmitter end.		dBm =	dBm
1.2 Minimal power requirement at receiver end			
- CNR requirement.	dB		
- Manufacturer's Specs for that CNR.		dBm	dBm
END-TO-END FIBER BUDGET		**1**	dBm

2. COMPUTATION OF THE LOSSES ON THE FIBER

2.1 Unallocated margin (e.g. 3 dB)	dB		dB
2.2 Optical fiber loss			
- Loss of chosen fiber		dB/Km	
- Average loss in splicing		dB/Km	
(e.g. Average no. of splices= 1 every 2 km			
Average loss per splice=0.2 dB)			
- Total loss per km	=	dB/Km	
- Line length	Km		
- Total loss in fiber link		x =	dB
2.3 Other passive device loss			
- Total number of connectors			
(Devices inputs/outputs, patch panels)			
- Average connector loss (e.g. 0.75 dB)		dB	
- Total connector loss		x =	dB
- Splitter, Isolator losses		dB	dB
2.4 Hazards margin			
- Aging	dB =		dB
- Temperature variation	dB =		dB
TOTAL LOSSES ON THE LINK		**2**	dB
REMAINING POWER MARGIN		**1** - **2** =	dB

Optical Link Budget Form

sensitivity to achieve a CNR of 51 dB is -34 dBm. Both of them do not consider any termination loss. These are estimated to 1 dB each.

Now we can calculate the link losses, to be compared to this budget:

• Safety Margin

It is better to decide at first upon this "unallocated" margin, which covers unavoidable components characteristics variations, and other malevolent effects (electronics aging is already taken into account in their power characteristics).

A safety margin of 3 dB is standard [63].

This standard should be applicated whenever possible. However, it may happen that special requirements drive the engineer to "squeeze out the last dB", and not to consider a sufficient safety margin. The link will still be safe to a certain extent, but it will be advisable to give special attention to the choice of its element and to its overall operation.

• Loss in Fiber

Typical loss per km of the fiber is given by the manufacturer. Additional loss is found in splicing. Each type of splicing has a maximum loss figure, and if the outside plant planning did not already give us an idea of the number of splices needed on the fiber, a splice every two km can be considered an average figure.
Splices as connections to splitters, for example, will be taken into account with the splitter loss.

Our 12-km single-mode link presents a typical attenuation of 0.4 dB/km; four splices are needed on the link, for 0.2 dB each.

• Loss in Connectors

Connectors located at the output of transmitter and input of receiver may already have been taken into account in the respective power characteristics.
Connector loss can have a range of values.

The chosen type of connector introduces a 0.75 dB loss. A patch panel is inserted at the transmission extremity of the link. With the connectors at transmission and reception ends, plus the connections to the coupler, the total number of connectors is five.

• Loss in Couplers

As described previously, the insertion loss is to be taken into account, plus the attenuation due to the connection splices on the fiber, when it is the case.

Here we decided to use connectors at each end of the coupler located at 6 km. This 50/50 splitter has a 4.1 dB insertion loss.

• Hazards Margin

Hazards margin takes care of components aging, temperature variation, etc. The 3 dB is a provision normally used in telephone company applications. It is not the same as the safety margin applicated at the beginning of our loss budget, though the same note applies here: provisioning a minimum hazards margin only diminishes the level of safety of the link and implies more attention to be given to every element to keep it in working order. (Note: it may be difficult in CATV applications to allow for a margin this large if the system gain is only 7 dB to start with; judgement needs to be exercised)

These give us a total loss on the link, and the difference with the acceptable loss calculated at first is the margin for changes allowed on the link.

The remaining power margin amounts to 4.9 dB. From this figure, we can now consider different possibilities of variations on the system.

Variations on the System

If the Remaining Power Margin is Positive

The system may still be able to work as is, if the input power does not exceed the dynamic range of the receiver, i.e. if it is not "overloaded", or there are a variety of choices on how to use the excess power:

• We can stretch the system to its maximum length.

Moving the receiving site away can be the best use of the remaining power margin, if the physical implementation is possible.

• We can save some fibers.

Typically at this level, the system format has been choosen a long time ago, mostly on a price basis. Signal quality is also a given. The calculation of the link budget allows us now to play with the number of channels we can transmit on a fiber, and still meet the needed performance.

• We can insert an attenuator, or save on the components price.

If loading the fibers to their maximum is not desirable, the whole link can be reconsidered, choosing lower quality components in terms of output power, input power of insertion loss. Another solution is to insert an attenuator, adjusted to meet the receptor dynamic range.

If the Remaining Power Margin is Negative

One solution is to relocate the receiver site to eliminate some loss on the link. Another is to opt for better quality components.

More drastically, we may have to reconsider the choice of the transmission format.

OPTICAL LINK BUDGET WORST CASE ANALYSIS	LINK DESIGNATION Tx End:___Example_ Rx End:_____ Number:_____	Nb of channels/fiber:_____ Nb of fiber links in the system:_____

1. END-TO-END FIBER BUDGET	ENGINEER'S CHOICES	MANUFACTURERS SPECS	POWER LEVELS
1.1 Guaranteed output power at transmitter end.		-10 dBm =	-10 dBm

1.2 Minimal power requirement
at receiver end
   - CNR requirement.
   - Manufacturer's Specs for that CNR.
	51 dB	-34 dBm →	-34 dBm

END-TO-END FIBER BUDGET		①	24 dBm
2. COMPUTATION OF THE LOSSES ON THE FIBER			
2.1 Unallocated margin (e.g. 3 dB)	3 dB →		3 dB

2.2 Optical fiber loss
   - Loss of chosen fiber
   - Average loss in splicing
   (e.g. Average no. of splices= 1 every 2 km
        Average loss per splice=0.2 dB)
   - Total loss per km
   - Line length
   - Total loss in fiber link
	+
	
	=
	12 Km
	 x
		0.4 dB/Km
		0.1 dB/Km
		
		0.5 dB/Km
		
		=
			
			6 dB

2.3 Other passive device loss
   - Total number of connectors
      (Devices inputs/outputs, patch panels)
   - Average connector loss (e.g. 0.75 dB)
   - Total connector loss
   - Splitter, Isolator losses
	4
	
	 x
		
		0.75 dB
		=
		4.1 dB →
			
			3 dB
			4.1 dB

2.4 Hazards margin
   - Aging
   - Temperature variation
	1.5 dB =
	1.5 dB =
			1.5 dB
			1.5 dB

TOTAL LOSSES ON THE LINK		②	19.1 dB
REMAINING POWER MARGIN		① - ② =	4.9 dB

Example of Optical Link Budget

4.6 Hometown Cablesystem Engineering

Let us return to the Hometown Cablesystem.

In this system, we would like to achieve, through fiber implementation:

> upgrade from 36 to 60 channels,
> consistent 50 dB signal to the subscriber,
> < 2 hours outage per year,
> preparation for future services.

Now is the time for the real decisions: we want to implement fiber tonight. What are the choices ?

Designing the Trunk

The characteristics of the trunk are:

> 20 km
> 60 channels
> 60 dB SNR at the hub
>
> $70 per subscriber for the total upgrade

Our attention will be directed at first to amplitude modulation, as being probably the least expensive option. Considering only the available alternatives (though new performance promises involving seductively small delays are appearing intensively in the papers) we find out that the nearest solution would be:

> 15 km
> 60 channels
> 57 dB
>
> $67 per subscriber, including coaxial distribution upgrade to 60 channels.

The 60 dB requested in signal quality cannot be achieved today using AM. In this example, we will have to modify our choice of format.

The next choice, according to the price, would be to use FM modulation on the supertrunk. We rapidly find that no manufacturer today is able to provide the type of service we expect so that we keep the total within $70 per subscriber, the equivalent of the AM microwave and coaxial upgrade.

Part of the budget allocated to performance upgrading will then be used on the supertrunk. We can try to get:

> 20 km (actually, up to 35 km)
> 60 channels
> 60 dB SNR
>
> $80 per subscriber

The distance margin would allow us to move the receiving site to a more convenient location, if desired. As it happens, we rented the building for the next 99 years, three years ago, and no moving around will be considered.

The supertrunk has now met our performance and economic aims. We will maintain the present solution for the moment.

Designing the Feeders

Pushing the Fiber to Six Subhubs

In Chapter 3, we decided on six subhubs, each serving 9000 subscribers. The requirements on these feeder links can be described as:

> 7 to 13 km
> 60 channels
> 55 dB SNR at the sub-hub
>
> $20 per subscriber (for the feeder portion only)

The economics did not change since step one, so: Is amplitude modulation able to meet these specifications ?

Survey is done, and so the requirements can be fulfilled for different prices, according to the length of the links:

— the two shortest links use a less expensive system providing 55 dB SNR up to 10km, with 18 channels per fiber,
— the four longer ones use a slightly more expensive system, up to 14 km, with 12 channels per fiber,

for a total of $14.1 per subscriber.

Though this figure fits into our budget, it can be interesting to study if the use of splitters is possible on certain links to see what the consequent savings may be.

Using our optical budget form, we rapidly find that the four longer links will not be able to support splitting of the signal.
On the two shortest ones (7 km each) the new optical budget, with a splitter located 3 km far from the hub, appears as in the figure titled Optical Link Budget for the First Hometown Feeder..

Some savings are generated here through the suppression of one transmitter and around 4 km of cable, and the total cost of applying fiber optics at the feeder level is now down to $13.34 per subscriber.

This is not a good example of the use of splitters in a network, since the tight loss budget drove us, as can be seen in the figure, to choose very high quality connectors (0.15 dB loss each) and splitters (0.25 dB excess loss), and be light on the different security margins. But with a lot of care in the installation, it would be operational.

Increasing the Number of Subhubs

Considering the difference between the $20 per subscriber we provided and the amount of money we actually spent in pushing the fiber down six feeders to 9000 subscriber areas, it is interesting to study the effect generated by the multiplication of SubHubs, i.e. pushing the fiber further, to even smaller areas.

Using the same approach, we find out that we could create nine subhubs, serving around 7000 subscribers each, for a total of $19.88 per subscriber, or $18.04 if two splitters are inserted near the Hub (here the splitters do introduce a bigger difference, the first feeders being shorter). This is close to our budget and fulfills the signal quality requirement, but does not leave a lot of space for future proofing or adjustments for enhancement of reliability.

These reasons drive us to choose the previous case, with six links and 9000-subscriber areas, as the most desirable at this point of the study. In future iterations we may wish to revisit this aspect of our system architecture.

OPTICAL LINK BUDGET WORST CASE ANALYSIS	LINK DESIGNATION	Hometown Cablesystem	
	Tx End:__ Hub___	Nb of channels/fiber:____ 18__	
	Rx End:__ SubHub1	Nb of fiber links in	
	Number:___ 1____	the system:____ 6____	

1. END-TO-END FIBER BUDGET	ENGINEER'S CHOICES	MANUFACTURERS SPECS	POWER LEVELS
1.1 Guaranteed output power at transmitter end.		-3 dBm =	-3 dBm
1.2 Minimal power requirement at receiver end			
- CNR requirement.	55 dB		
- Manufacturer's Specs for that CNR.		-10.5 dBm ←	-10.5 dBm
END-TO-END FIBER BUDGET		①	7.5 dBm

2. COMPUTATION OF THE LOSSES ON THE FIBER

	ENGINEER'S CHOICES	MANUFACTURERS SPECS	POWER LEVELS
2.1 Unallocated margin (e.g. 3 dB)	0 dB →		0 dB
2.2 Optical fiber loss			
- Loss of chosen fiber		0.4 dB/Km +	
- Average loss in splicing		0.1 dB/Km	
(e.g. Average nb of splices= 1 every 2 km Average loss per splice=0.2 dB)			
- Total loss per km		= 0.5 dB/Km	
- Line length	7 Km		
- Total loss in fiber link		x =	3.5 dB
2.3 Other passive device loss			
- Total number of connectors (Devices inputs/outputs, patch panels)	2		
- Average connector loss (e.g. 0.75 dB)		0.15 dB	
- Total connector loss		x =	0.30 dB
- Splitter, Isolator losses		3.25 dB →	3.25 dB
2.4 Hazards margin			
- Aging	0.2 dB	=	0.2 dB
- Temperature variation	0.25 dB	=	0.25 dB
TOTAL LOSSES ON THE LINK		②	7.5 dB
REMAINING POWER MARGIN		① - ② =	dB

Optical Link Budget for the First Hometown Feeder

References

1. Optical Fiber Communication Conference (OFC 89), *1989 Technical Digest Series*, Vol. 5 (Optical Society of America, Washington, D.C., 1988), and recent articles in different publications including Lightwave Journal, February 89, pg 1.

2. Jenkins, F., *Fundamentals of Optics*, Fourth Edition, McGraw-Hill Inc., 1976, p. 19.

3. Quoted by Cabletel of Montréal, Canada, in a seminar on fiber optics, 1988

4. The wavelength (i.e. the "color") of the light being used for transmission is here expressed in nanometers (1 nm = 0.000000001 meter).

5. Nérou, J.P., *Les Fibres Optiques*, Le Griffon d'Argile, Inc., Québec, 1983, pp. 55, 131

6. Spectral width: we all know at least one "spectrum" of light: the colors of the rainbow. Fiber optics also has a spectrum, but the objective is to transmit only one frequency (i.e. one "color") of the light. Thus "spectral width" defines the variation in the colour of the light. The "narrower" the spectral width, the more pure the light and the better for optical transmission.

7. Numerical aperture of a fiber: the acceptance angle of a fiber is the angle under which light will be "accepted" into the fiber. The numerical aperture is defined as the sine of half of the acceptance angle of the fiber, in degrees; i.e. it is a number defining the "opening" available for the light.

8. Baker, D.G., *Monomode Fiber-Optic Design*, Van Nostrand Reinhold, 1987, p.142.

9. Information received from Gould Electronics, Inc., 1989

10. Information received from Canstar, Inc., 1989

11. From Gould Electronics, in terms of excess loss from 0.75 dB to 0.25 dB and then to 0.07 dB, in terms of coupling ratio tolerance from 10% to 5% and then to 3%, in terms of polarization stability from 2% to 1.5% and then to 0.5% will double, and then quadruple the price of the single-mode coupler.

12. Information received from JDS Optics, 1989

13. Information received from Isowave, Inc., 1989

14. J.Holobinko, American Lightwave Systems, quoted 85% for transmission site and 15% for reception site in analog transmission, in seminar ADCOM Electronics, Ltd, Toronto, March

7, 1989.

15.Electroluminescent materials emit light when stimulated by electricity.

16. Price is an issue today, but a laser has much the same structure as an edge-emitting LED. As the technology matures this price difference should become less important.

17. Baker, D.G., op. cit., p.303.

18. Lifetime of a laser can be quoted as its MTBF (mean time before failure).

19. Information received from Orchard Communications, Inc.

20. Several quotations give different figures, for example J.Holobinko, ALS, quotes 200,000 hr, meaning 23 years.

21. Among other references, FP lasers are quoted as one-third less expensive than DFB lasers in Luff, R.A., from Jones Intercable, The Broward Cable Area Network Fiber Model, *Communications Engineering and Design*, February 1989, p. 27.

22. Lightwave Journal, Janurary 1989.

23. Baker, D.G., op. cit., p.305.

24. Nérou, J.P., op. cit.

25. Integrated detector-preamplifier provide a detector and a preamp on the same integrated circuit. The noise figure is improved, and so is the signal quality.

26. Nérou, J.P., op. cit., p.131.

27. Information received from Ericsson Telecom, 1989.

28. Baker, D.G., op. cit., p.155

29. Thylen, L., Optical Switch Arrays in LiNbO3, Status Review and Prospects, Integrated and Guided-Wave Optics, *1989 Technical Digest Series*, Vol. 4, (Optical Society of America, Washington, D.C., 1989), p.2.

30. DiCon Fiber Optics, Inc. advertises typical insertion losses of 0.5 dB for multimode fibers and 1.2 dB for single-mode fibers with fiber moving technology, up to a 2x2 configuration. Maximum dimension is 2x39 fibers.

31. Strictly Non-blocking 8x8 Integrated Optical Switch Matrix, *Electronic Letters*, Vol.22

(1986), p. 816, referenced by Catel, quotes an example of performance: Insertion loss 4-5 dB, crosstalk approximatively -30 dB.

32. Olsson, N., Optical Amplifiers, Optical Fiber Communication Conference (OFC 89), *1989 Tutorial Papers*, Vol. 5 (Optical Society of America, Washington, DC 1988), p. 343.

33. Information from British Telecom, UK.

34. Way, W., Subcarrier Multiplexed Lightwave Systems, Optical Fiber Communication Conference (OFC 89), *1989 Tutorial Papers*, Vol. 5 (Optical Society of America, Washington, D.C., 1988), p.55.

35. Baker, D.G., op. cit., p.361.

36. Gysel, H., Composite Triple Beat and Noise in a Fiber Optic Link using Laser Diodes,Convention and Exposition of the National Cable Television Association (NCTA), *1988 Technical Papers*, p.94. See also: Properties and Systems Calculations of Optical Super-trunks for Multichannel TV Transmission, using Analog Intensity Modulation, Single-mode Fibers and High Deviation FM, ibid.

37. Baker, D.G., op. cit., p.309.

38. Gysel, H., op. cit.

39. The video SNR is a linear function of CNR above the FM detector threshold. Various relations exist, depending on EIA or CCIR weighting, 100 or 140 IRE reference.

Orchard Communications provided us with the following:
SNR = CNR(4.2) + 12 + 20 log(D_{STPW})
Where D_{STPW} is the deviation of the non-emphasized part of the video signal: it is called "Sync tip to peak white" deviation.

40. Most quoted are the variations of the signal quality at a given loss budget, derived from practical measures. For example:

• From American Lightwave Systems, on FM:

CNR(N) = CNR(1) -20 log N   for 1 < N < 3
CNR(N) = CNR(1) -15 log N   for 4 < N < 19
CNR(N) = CNR(1) -10 log N   for 20 < N < 40

• From Anixter, on AM:

CNR(N) = CNR(1) -10 log N   for 10 < N < 50

41. Adams, M., Composite Second Order: fact or fantasy?, *Communications Engineering and Design*, November 1988.

42. Gysel, H., op. cit.

43. Information received from ALS.

44. Information received from Anixter.

45. Information received from Catel.

46. Gysel, H., op. cit.

47. Information received from Comlux.

48. Holobinko, J., American Lightwave Systems, in seminar ADCOM, Toronto, March 7, 1989.

49. Lightwave Journal , March 1989.

50. Information received from Anixter.

51. "Independantly of the manufacturer, DFB lasers provide 42 channels per fiber, 18 km reach, 51 dB SNR, and sound like the right approach to AM [transmission]", in *Lightwave Journal*, January 1989.

52. Gysel, H., op. cit.

53. Appearing among others in *Lightwave Journal* , February 1989

54. Baker, D.G., op. cit.

55. Information received from Catel.

56. To minimize the decrease in signal quality, drop and insert are performed without demodulating the signal in transceivers that work as repeaters. (Information from Orchard Communications & Foundation Instruments.)

57. Some methods to handle scrambling have the back clamp replaced with the average content of each line .

58. The Zenith system, a popular CATV scrambling scheme, apparently cannot be transmitted on any FM system.

59. Holobinko, J., cit.

60. Information received from Comlux.

61. Information received from Comlux.

62. This is the linear worst-case approach, where each element contributes with its worse possible characteristic. Another possible approach is a statistical method, claimed to give more precise analysis, but much more difficult to implement. Here we will keep the security of the worst-case analysis.

63. Holobinko, J., cit.

Chapter 5
Construction and Maintenance

5.1 Optical Fiber and Cable Design Characteristics

Optical Fiber

Optical fiber is a medium which uses light to transmit information. Compared to a coaxial cable or a pair of copper wires, optical fiber has a very small diameter: about .1 mm.

Mechanically, fiber is very resistant to tension. However, its transmission properties and life span could be affected by application of constant tensile stress.

On the other hand, its bending strength is relatively poor. Even with the highest manufacturing standards currently in use, the minimum bend radius is about 500 times the fiber's diameter. A tight bend will affect the fiber's transmission properties and life span.

The fiber itself can be affected by a number of environmental stresses. For instance, water in all of its forms (liquid, vapor, moisture, mold, ice, etc.) attacks fiber in a process called hydrogenation which can create microcracks in the fiber and which may ultimately result in fiber failure.

Optical Fiber Cables

When installing optical cables, there are two main areas to be considered:

• The type of fiber configuration within the cable;

• The cable design.

Cable Configuration

There are four main configurations for cabling fibers:

• Loose tube buffer;

• Tight buffer;

• Open slot or star tube;

• Ribbon.

The figure, "Configurations of Optical Fiber Cables", shows a cross section of each type of configuration.

In a loose tube design, the fibers are subject to very little tensile stress, as they are independent of the actual cable. When the cable is subjected to stress, during installation for instance, the strength member and the sheath absorb the tension, so that the fibers are not affected. Also, when the cable contracts or stretches due to changes in temperature, the fibers remain loose and unaffected.

In a tight buffer type of design, the fibers undergo the same tensile stress as the cable does. On the other hand, if the cable breaks, the fibers all break at the same place. A fiber will rarely break at a distance of several meters away from the point at which the cable broke. This facilitates the identification of the repair location, and admits a simple splicing repair instead of changing a part of the cable.

An open slot or star tube type of design is similar to the loose tube buffer, the main difference being that the fibers are all found at the surface of the cable, closer to the cable's exterior, depending on the composition and thickness of the cable's sheath.

Finally, with the ribbon cable there can be a greater fiber count within a given diameter, but the advantages and drawbacks are the same as with the tight buffer.

Loose Tube Buffer

Central member

Fiber

Loose tube buffer

Interstitial filling

Sheath (Kevlar)

Polyethylene jacket

Tight Buffer

Central member

Fiber

Interstitial filling

Polyethylene jacket

Open Slot

Fiber

Extruded plastic

Polyethylene jacket

Heat barrier

Central member

Ribbon Cable

Strength member

Fibers ribbon

Metallic tape

Thermal tape

Interstitial filling

Fiber

Fibers ribbon

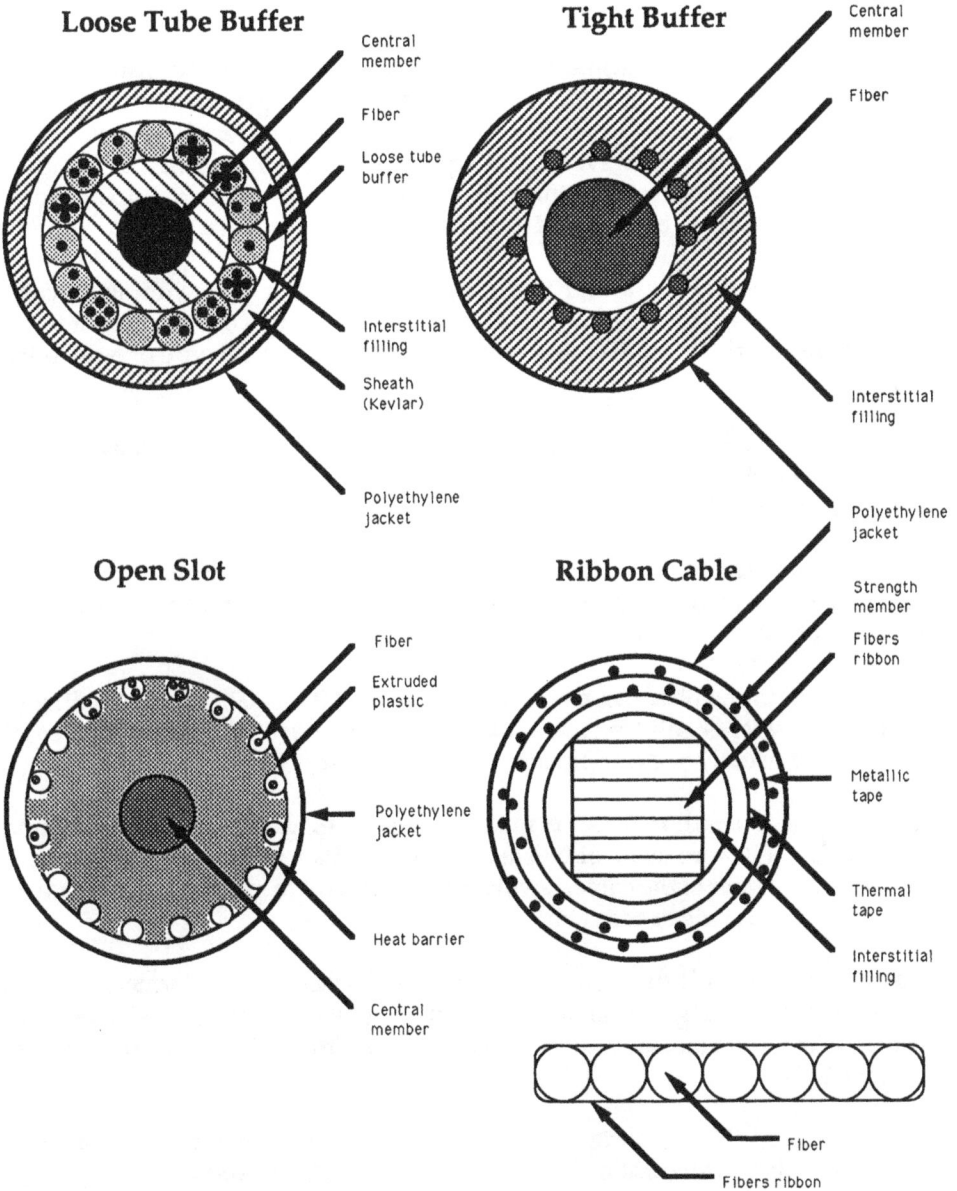

Configurations of Optical Fiber Cables

Cable Characteristics

As with coaxial cables to some extent, fiber optic cables may be installed with various outer layers. Fiber optic cables also have a strength member which can withstand the stresses of installation and environment, especially tensile stress. This strength member, however, is not equivalent to the strand that is found on some coaxial cables (self-supporting cable) and which acts as cable support. Some fiber optic cables may be supplied as self supporting.

The strength member is found under many different forms: with some cable designs, it occupies the center of the cable, while with other constructions it forms a protective sheath. It can be made of solid steel, twisted steel wire or a dielectric material.

If the strength member is found at the center of the cable, the fibers may be more easily accessed than if it were on the outside. On the other hand, an outer member will better protect the fibers against environmental stresses.

If the strength member is made of steel, the cable must be grounded regularly in order to avoid any tension differences which could occur between the cable and any equipment connected to it.

A steel strength member has a tensile strength of 600 to 1,000 lbs, depending on whether the steel is solid or made of twisted steel wire. A dielectric strength member will usually provide smaller tensile strength.

Finally, if the strength member is made of solid steel, grounding becomes more difficult than if it were made of twisted steel wire. Also, if the strength member is located at the center of the cable, it will be more difficult to ground it than if it were located on the outside of the cable.

Sheaths are available in different configurations. They are usually formed of several layers, each protecting the cable from a certain type of stress. There is usually a water and moisture blocking layer, a layer to protect the cable against rodents, and an outer jacket.

The tables titled "Fiber Configurations" outline the advantages and drawbacks of the various fiber configurations and cable design characteristics.

TABLE OF FIBER OPTIC CABLE CONFIGURATIONS

STRESSES	LOOSE TUBE	TIGHT BUFFER	OPEN SLOT	RIBBON CABLE
Tensile	None	All	None	All
Bending	None	All	None	All
Fibers break with cable	Cable breaks over several meters	At the point where the cable breaks	Usually close to or at cable's breaking point	At the cable's breaking point

TABLE OF FIBER OPTIC CABLE CHARACTERISTICS - 1

STRENGTH MEMBER	INTERNAL	EXTERNAL	SOLID STEEL	TWISTED STEEL WIRE	DIELEC-TRIC
Access to fibers	Easy	Difficult			
Grounding			Required Difficult	Required Easy	Not required
Tensile strength			600 lbs	1,000 lbs	Low
Cost			Low cost	Costly	Costly

TABLE OF FIBER OPTIC CABLE CHARACTERISTICS - 2

PROTEC-TIVE COM-POUNDS	WATER	MOISTURE	RODENTS	AIR	ELECTRI-CITY	UV RAYS
Gel fill	Excellent	Good	Poor	Fair	Poor	Poor
Water-blocking tape	Excellent	Excellent	Poor	Good	Good	Good
Metallic tape	Fair	Fair	Excellent	Fair	Bad	Fair

5.2 Building a Fiber Optic Cable System

Even though the installation of fiber optic cables is quite similar to that of coaxial cables, planning construction of a cable route is quite different.

Coaxial cables are delivered in rolls of about 750 m each, while rolls of fiber optic cables may reach 6,000 m or more.

One reason why rolls of coaxial cable are so short compared to their fiber optic counterpart is the need for amplifiers at intervals of less than 750 m.

Since optical fiber is available in very long rolls and amplifiers are not required, the installer must carefully plan his cable run in order to reduce the number of obstacles to be bypassed and to keep the number of splices at a minimum.

For instance, with an aerial installation covering about 5 km, how many times does the fiber optic cable have to be installed above existing strands? What method should be used to pass the fiber optic cable above other cables? At which point in the run should the cable be back pulled?

Also, in duct installations, if the splices are usually placed at almost every manhole in a coaxial cable network, the direction to be taken and the starting point become very important when installing the optical fiber.

For underground installations, the same question arises concerning where to begin the run to minimize the number of splices. In this instance, splices will be set according to the number of detours to be made.

In a coaxial cable network, since there is a large number of connectors per km (dividers, amplifiers, power injectors, etc.), and since these connectors or splices do not generate substantial loss, hardly any attention is brought to a cable's splicing point. In fiber optics, a splice corresponds to a loss generated by several hundred meters of cable, which is quite substantial.

Therefore, splice loss being relatively important, the cable run should be carefully planned in order to bypass areas where the cable has a higher risk of being damaged and, as a result, additional splices would be required.

Additional Fiber Optic Hardware

There is almost never a need for amplifiers on an optical fiber link. Amplifier-repeaters, therefore, will not be considered here. However, following an accident, there may be a need for splices on the cable. Connectors at each end of the cable serve to connect the cable to active or passive hardware. The following section will deal with the various types of fiber optic hardware.

Connectors

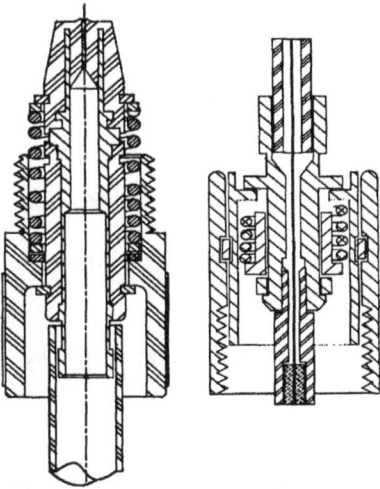

Examples of Connectors:
Biconic and D4

(After: *Fiber Optic Technology,*
Document AMP Inc., 1988.)

There are many different types of fiber optic connectors. The "Fiberoptic Product News, 1988/89 Buying Guide" listed 50 manufacturers, with 224 connector model numbers.

Some connectors can be easily connected to the fiber by the user, while others must be assembled in the plant and are sent in the form of pigtails. They must then be connected to the fiber through a splice.

More often than not, a connector will be chosen according to the type of connection required between the fiber and the hardware.

We will consider here four of the main types of connectors: Biconic, D4, FC and S. Each type has its own set of characteristics and prices.

Until recently, most optical connectors did not allow any physical contact between the fiber and the piece of hardware. This resulted in insertion loss and substantial reflection. However, the new series of PC (Physical Contact) connectors have considerably improved connector performance.

CHARACTERISTICS OF CONNECTORS		
TYPE	REFLECTION LOSS (dB)	INSTALLATION (minutes)
BICONIC		15
D4	-45	
FC	-45	
ST	-45	16

Splices

There are two types of optical fiber splices: mechanical splices and fusion splices.

The first type of splice mechanically aligns two fibers from one end to the other. Once aligned, the fibers are held into place with epoxy glue.

Even though they need special handling, mechanical splices are relatively easy to set and require only a few tools. However, they are inferior in many respects to fusion splices.

Though most of these splices consist of tapered tubes which contain aligned fibers, there is a type of mechanical splice which consists simply of aligning the fibers over a gummed ribbon enclosed in a small plastic box.

Fusion splices require more sophisticated hardware and more highly skilled workers, and involve an upfront investment in fusion set of $35,000-$100,000 or so [1]. However, their performance level is superior to that of mechanical splices.

Fusion splicing consists in aligning the fibers either manually or automatically, depending on the mechanism used, so that insertion losses are minimized. Then, both fibers are fused together usually through heat emitted by a laser beam.

A mechanical splice will usually generate losses between .5 dB and 1 dB, and

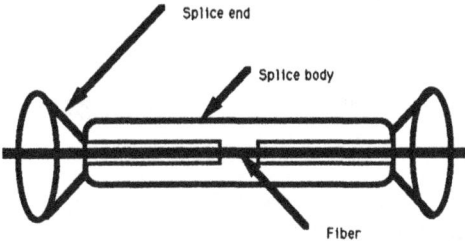

Regular Mechanical Splice

sometimes even more, whereas losses resulting from a fusion splice can be as low as .1 dB. As for reflection performance, field trials showed fusion splices typically exceeding -50 dB, while mechanical splices needed careful manipulation to achieve -40 dB [2].

Therefore, fusion splices should always be used when either a link is being installed and permanent repairs are being done. The use of mechanical splices should be limited to temporarily restoring a defective link.

Splice Cases

Fiber optic splices must be installed in splice cases, which usually consist of a splice organizer and a splice tray.

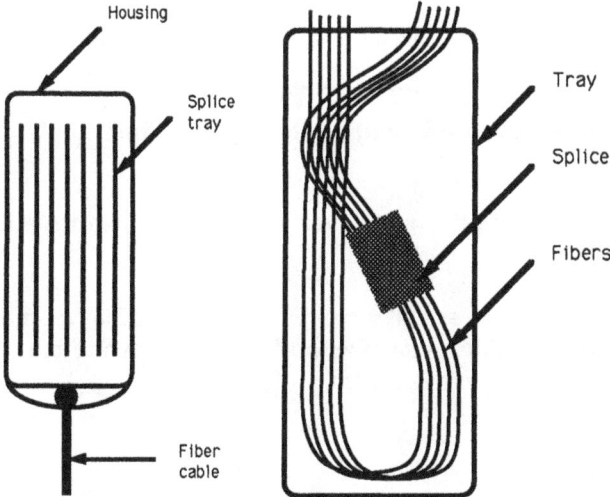

Splice Closure and Splice Tray

The case must be designed with the bending characteristics of the fiber in mind. It should also be completely waterproof, so that the splices will be protected from the environment.

Patch Panels

Though not mandatory, patch panels are an efficient means of connecting cables with hardware.

Thus, cables may be permanently fastened to panels while patch cords link up with hardware. If a piece of hardware must be changed or moved around, only patch cord connectors are disconnected or must be changed so that the cord is long enough for the new location.

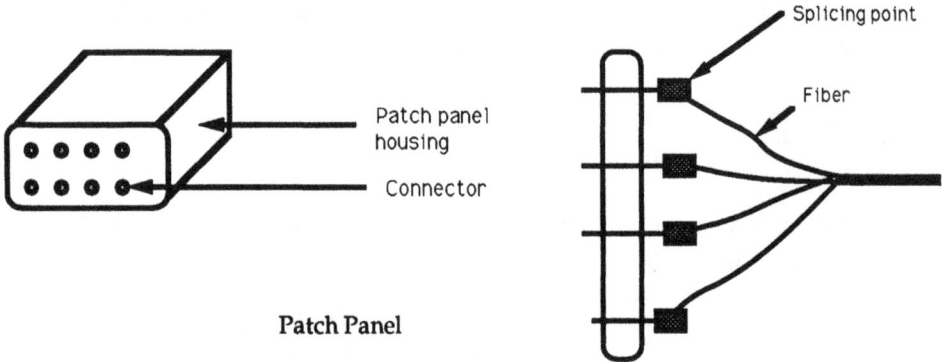

Patch panel housing

Connector

Splicing point

Fiber

Patch Panel

Planning a Cable Run and Installing the Optical Fiber

Carefully planning the cable run is very important when installing optical fiber cable.

It is important to remember that a reel of optical cable may be many times longer than a reel of coaxial cable. Also, fiber optic splices generate substantial losses and care must be taken to minimize them.

Aerial Installation

All of the usual cable installation methods may be used to install optical fiber.

A fiber optic cable may be lashed to existing cable or it may be placed on a new strand. Ideally, the cable should be lashed to the highest strand to minimize expo-

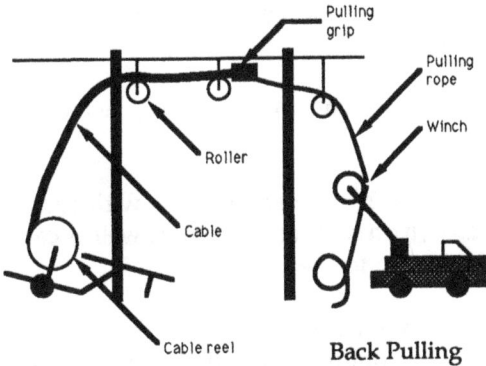

Pulling grip

Pulling rope

Winch

Roller

Cable

Cable reel

Back Pulling

sure when work is being performed on other cables supported by the same pole.

The cable run must be carefully planned. The following should be taken into account:

• Crossing above existing cables and strands;

• Crossing with branch lines and drops;

• Trees with branches lower than the strand on which the fiber will be placed;

• Underground street crossings.

Once these factors have been carefully analyzed, it can then be decided whether the cable should be lashed directly without first being stretched, at which points should the cable first be stretched out, and where in the run should the installation begin.

It will then be possible to determine the points at which cable pullers and guides will have to be installed, to better support the cable when it is pulled.

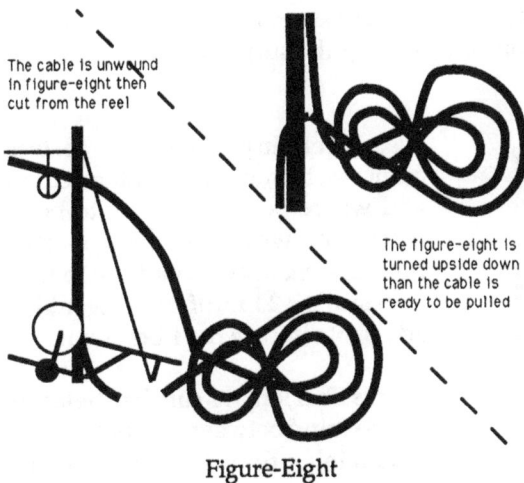

The cable is unwound in figure-eight then cut from the reel

The figure-eight is turned upside down than the cable is ready to be pulled

Figure-Eight

Should the cable have to be pulled at several points, it may then be laid out in a figure-eight configuration at a given point, so that enough cable will be retained to reach the second point, before continuing the pulling process.

When there are no obstacles along the route, the cable may be unwound and lashed at the same time, using standard methods.

Whatever the technique used, the following main points should be carefully considered:

• The tensile load strength recommended by the cable manufacturer should be adhered to so that the fibers will not be damaged.

• The manufacturer's recommended minimum bend radius should not be exceeded to avoid damaging fibers.

During installation, expansion loops should be left at each pole as with coaxial cables, in order to minimize the pulling tension on cable and fibers, as well as cable and strand contraction resulting from changes in temperature.

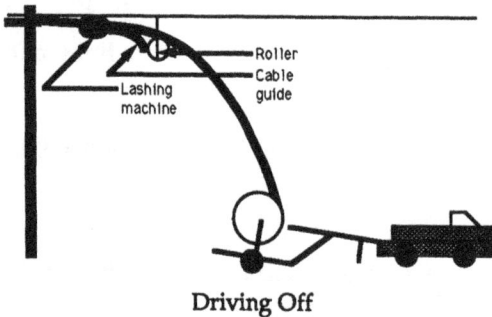

Driving Off

A good idea is to plan first the exact location of the splices. There should be as little equipment as possible (e.g. electrical transformers, telephone company loading coils, amplifiers or cable network power supply) at the pole next to which the splice is to be located in order to minimize potential exposure to the splice.

When installing the cable, 10-15 m of cable should be left at the splicing point so that splicing may be performed on the ground rather than in a pole; also, surplus cable will be required for any new splice, and in case the cable should have to be moved temporarily, i.e. if an elevated structure has to pass under the cables, or to avoid having to perform two splices if the fiber should break. (Fibers may break at intervals of several meters, which would require replacing a few meters of cable or else using the surplus cable left initially for that purpose.)

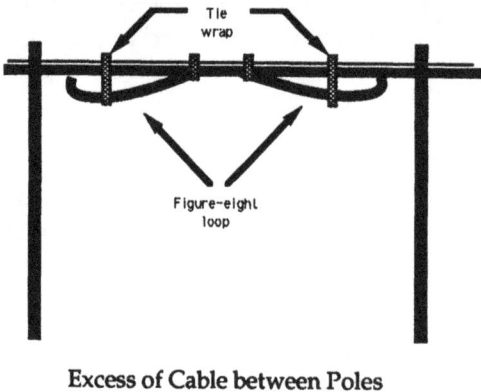

Excess of Cable between Poles

Also, in certain locations such as street corners which are major intersections, and where poles are likely to be replaced, or where electrical installations are changed from time to time, to retain 15-20 m of cable would be advisable in case the fiber is cut.

Excess fiber lengths can then be lashed to the strand between two poles, while maintaining the cable's bend radius.

Duct Installation

Installing a cable which is several kilometers long can be a complicated process. There are several manholes to be crossed, and these are often cluttered. Also, the ducts are not always in good condition, the conduit structures are often winding, and this can result in additional friction which makes pulling the cable difficult.

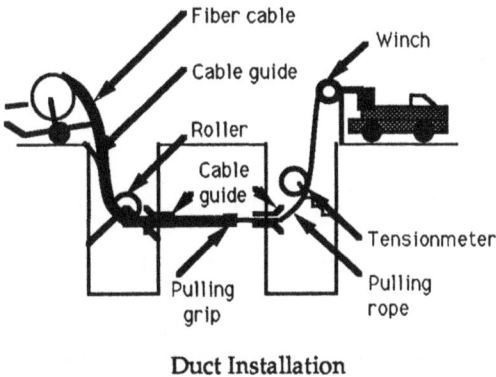

Fiber cable

Winch

Cable guide

Roller

Cable guide

Tensionmeter

Pulling grip

Pulling rope

Duct Installation

By analyzing the structure and its state, it is possible to determine the best starting point for the cable run. From there, the cable may be pulled one way, while a sufficient length of cable is then unwound to reach the other end before it is pulled in the ducts.

When installing cable in a duct, there should be one person in each manhole to keep track of the procedure; also, cable guides should be placed at both ends of the duct so that the cable may travel along the walls of the manhole.

During the pulling process, the manholes should be equipped with tensionmeters to measure the tension placed on the cable so that the manufacturing standards are never exceeded.

Lubricants may be used to reduce tension, in the same manner as with coaxial cables.

If possible, an empty duct should be used to install fiber optic cables. The cable may be installed in a duct which is already being used if the tensile stress load is constantly watched. However, installing new cables in a duct is not recommended where there are already one or more fiber optic cables.

The cable must be carefully placed along the walls inside the manhole and, if possible, it should be covered with protective materials to avoid damage during construction work in the manhole.

Whenever possible, several meters of surplus cable should be kept in the manhole for future needs. This length of cable can be easily attached to the manhole walls.

Buried Cable Installation

Once again, the cable run must be carefully planned.

As for coaxial, the cable should be placed at a depth of 100-125 cm so that it will not be affected by shifts in the earth resulting from thawing and freezing, or other moves depending on the climate. Installing the cable at such a depth also has the following advantages:

• Fewer rodents;

• Less danger from water seepage;

• Less pressure resulting from surface conditions (e.g. heavy vehicle traffic);

•Lower chance of damage to fibers resulting from construction work.

Should the cable be installed in an open trench, there should be a layer of sand or gravel with the cable placed on top. If the area is potentially hazardous, the cable may also be covered in concrete. When a cable is placed in a trench, care must be taken when filling the trench to avoid that heavy rocks lean against the cable and damage it.

In all underground installations, it is advisable to place a warning tape a few centimeters above the cable to avoid potential damage to the cable during excavation work. Also, every underground installation should be marked regularly.

Splices may be put in cases and placed in trenches, but it is better to put them above ground, on pedestals, where they can be easily located and accessed. A buried splice should always be identified by a marker.

Installation Costs

The cost of installing fiber optic cable, just like any other cable, depends on a great many factors. The prices that are given here are approximate, and serve only to provide an overview of construction costs [3].

For instance, the cost of installing a cable with over 24 fibers, lashed onto existing cables, may vary from less than $2(CDN), to more than $5(CDN)/m.

Installing the same cable on a naked strand would cost around $0.50(CDN) less per meter. However, the cost of installing this cable in underground ducts could fluctuate from $3(CDN)/m to $10(CDN)/m. Finally, if the above cable must be buried, the cost of installing it could go from $3(CDN)/m to $10(CDN)/m.

5.3 Environmental Stresses

Despite all the means used to protect fiber optic cable and splices, a number of environmental stresses can in fact reduce their life span or create system failures which take a certain time to repair.

These stresses may be natural or induced by man [4]. Rain and electrical storms, for instance, are natural stresses, while a pole which has been hit by a car is a man-induced stress.Even if one cannot possibly anticipate every stress imaginable, there are ways to minimize the effect of most of the stresses to which the cable may be subjected.

Natural and man-induced stresses are depicted in the two tables that follow.

Some solutions are more efficient than others when trying to pinpoint potential natural and man-induced stresses to the cable. However, though much care may be taken in planning and installing the cable, a catastrophe is almost inevitable. Nevertheless, the more carefully a cable run is planned, the less damage will result. Therefore, when installing a fiber optic line, the first thing to keep in mind is that the runs must be properly planned, and the cable carefully chosen.

Electronic Equipment

Just like many other highly sophisticated equipment, fiber optic hardware is quite sensitive to environmental conditions. Transmission hardware should be operated in atmosphere-controlled facilities, as is most other equipment of a similar nature.

Some types of hardware — most AM optical receivers, for instance — have been designed to be kept outside and are thus resistant to extremes in climate. However, even though they may be relatively resistant to the natural environmental stresses

TABLE OF NATURAL STRESSES ON CABLES

STRESS	EFFECTS ON THE CABLE	SOLUTION
Changes in Temperature	Stretching and contracting	Lash all similar types of cables together
Freezing and Thawing	Compression (underground or duct) by ice	Bury cables deep enough to avoid freezing
	Shearing due to earth shifts resulting from freezing and thawing	
Wind	Weakening of members and sheaths	Choose materials which are wind-resistant
	Tensile stress on splices	Place splices away from vibration
Rain and Moisture	Corroded metallic sheaths	Cover sheaths with waterproofing compounds
	Water seeping into cable resulting in fiber hydrogenation	Use waterproof gel inside cables
Snow, Ice and Frozen Rain	Overloading strands which may lead to sagging	Calculate the strand's diameter to check if it can support all of the cables lashed on as well as anticipated amounts of ice and frozen rain
Electrical Storm	Splits open the cables' metallic sheaths	Cover the cable with an insulating material
Insects and Rodents	Chewing	Use metallic sheaths

TABLE OF MAN-INDUCED STRESSES ON CABLES

STRESS	EFFECTS ON THE CABLE	SOLUTION
Traffic Accidents	Stretching of the cables	Use stress-resistant strands at potentially hazardous areas
		Place splices at safer points
	Overtensioning in the splices	Lift up strands to ensure that there is minimum safety clearance
	Cable and splice breaking	Cut across hazardous areas in underground ducts covered in concrete
Excavation	Cable breaking	Properly mark off underground cables
		Place warning tapes above underground cables
		Use ducts covered in concrete for the most hazardous areas

previously mentioned, man-induced stresses must be considered when deciding where to place such equipment.

5.4 Maintenance and Upkeep

The maintenance and upkeep of a fiber optic network is the same as that of any CATV network — it begins once the network becomes operational. In fact, the initial surveys will form the basis of a preventive maintenance program and will enable the system to be followed up. It would be advisable, from the very first splice, to make note of the first optical network performance surveys.

Before the first splice is done, the fibers should be tested to ensure that their attenuation meets manufacturing standards. If such is not the case, then the fibers were most probably damaged during installation. An optical reflectometer should be able to indicate where the breaks, if any, are located.

Once a splice has been set, the cable can be tested in order to gauge actual splice losses and the splice's level of reflection.

An example of the "Fiber Optic Link Survey" form which follows can be used as a follow-up on an optical link.

From time to time, measurements should be made on each optical link. These measurement results should then be compared to previous ones, as well as to the initial measurements of the optical link.

These comparative measurements allow a follow-up to be done on the link and any weaknesses may then be detected so that potential problems can be pinpointed before a major link failure occurs.

Testing Equipment [5]

There are many kinds of testing equipment both useful and necessary in calibrating and maintaining an optical network.

The most basic test set measures the signal's intensity. It is equivalent to a field strength meter. Every company which operates a fiber optic system should have at least one of them. The cost for this ranges from several hundred dollars to about $4,500.

The second type of testing equipment which is useful is the optical reflectometer. It performs the same set of tests within the optical bands as the RF reflectometer does in its frequency band. It measures the reflections, locates the splices and cable breaks, etc. An optical reflectometer can cost anywhere from $8,000 to $40,000, depending on how sophisticated and precise it is.

Finally, the dispersion/bandwidth tester is possibly the most advanced apparatus on the market. It is a type of sophisticated optical reflectometer. It costs between $15,000 and $125,000. It is not absolutely essential to an optical network operator, especially since a good reflectometer provides all the necessary measurements.

FIBER OPTIC LINK SURVEY

Link Identification: _____

	Fiber Id.	Length	Attenuation	Splice # Atten. Refl.		Total Atten. Refl.	
Theoretical Calculation							
Total							
Installation							
Total							
Survey Date							
Total							
Survey Date							
Total							

Link Survey Form

Repair Equipment

In repairing optical system failures, time is of the essence. Whether repairing transmitter or receiver hardware, or damaged fiber, repair time must be minimal.

Though a coaxial network usually contains a number of replacement parts for all of the system's hardware, such is not necessarily the case with an optical network.

The spare parts inventory should not generally exceed 2-3% of the total hardware. On the other hand, what if a transmitter should break down in a system of only four transmitters?

A signal power surplus may possibly be available on a link and it may be able to power the link when there was a transmitter breakdown. In a multi-link system, for instance, several links may have to be transferred to other transmitters in order to power a link where the transmitter broke down. Careful planning will enable any transfers to be done quickly, should a transmitter break down.

As for optical receivers, since they usually outnumber transmitters, it would be wise to have an additional receiver should there be a need for it. The most exposed element in an optical network is without doubt the optical fiber itself. Therefore, repair work will mostly deal with fiber optic failure.

Mechanical tape splices are available at low cost, and they are easy to install, but relatively inefficient. However, they are useful in repairing a cable quickly. Fiber optic users should have at least a few tape splices, and each technician repairing cables should ideally have a tape splice kit.

Conventional mechanical splices give a best result. Though the repair work is perhaps only temporary, the service is re-established quickly. A mechanical splice kit contains all of the necessary tools to perform a splice. Fiber optic users should all have at least one such kit.

Finally, fusion splicing should always take the place of mechanical splicing when final repairs are being done. However, the required hardware is costly and may appear too complicated to the layman.

A fusion splicing machine is not necessarily required for small systems. Specialized contractors may be called upon to effect permanent repairs when required, if the user is able to temporarily repair a broken cable using mechanical splicing.

5.5 Hometown Cablesystem Construction and Maintenance

This section presents Hometown construction and maintenance activities which hypothetically occurred earlier this year in implementing the strategic plan. Here we will look at a rebuilt portion of this network which is served by a sole microwave receiver.

The receiver is located at a hub site about 20 km from the head-end. The sector being served has been divided into seven subsectors (six subhubs plus the hub), each subhub sector having a coaxial distribution network consisting of up to six or seven cascaded amplifiers.

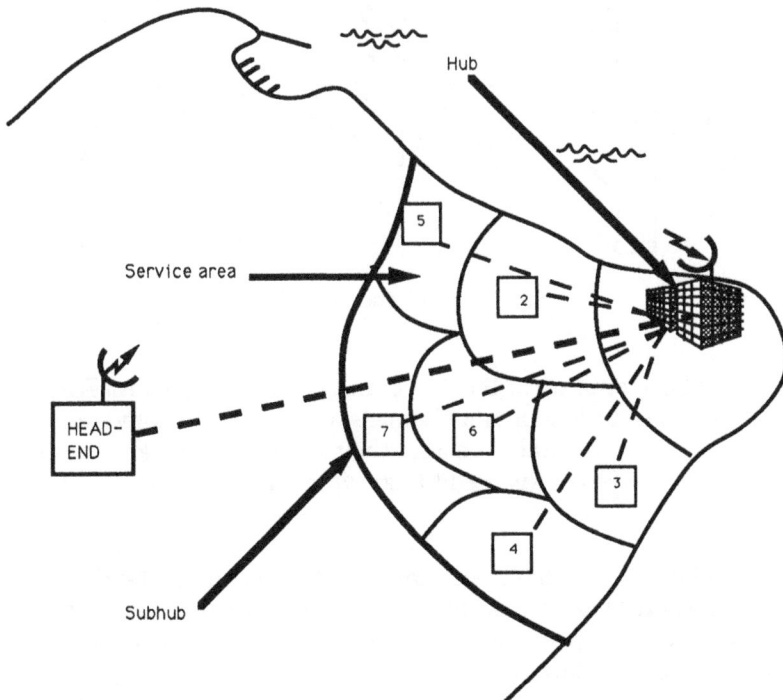

Overview of the Hometown Cablesystem Fiber Network

Electronic Equipment

The head-end and the hub will be located where the current microwave hardware is. The microwave link can serve as an emergency power source should the optical link fail. These installations are also already equipped with emergency power sources to back up the system in case of power failures, with cooling systems and with equipments of protection against fire, theft, or vandalism.

At the head-end, all FM and optical transmission hardware were placed in additional equipment shelves.

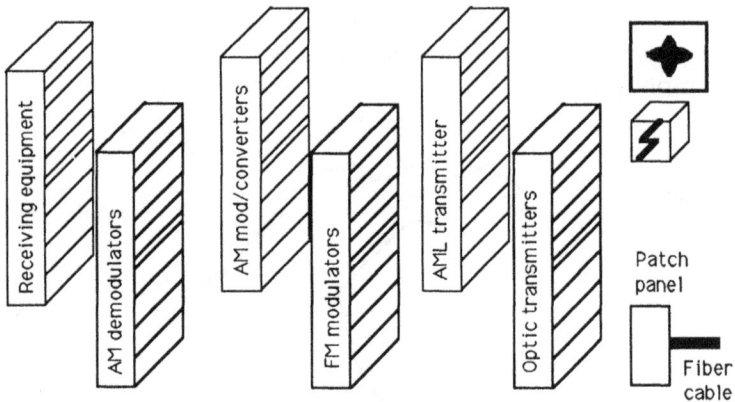

Head-end Organization

The signal has been split up at the AM demodulator outlets, and additional demodulators have been added for the channels which are to be converted to another frequency, or which are retransmitted to their receiving frequency. Thus, all the channels were routed towards the FM hardware outlets in order to be remodulated.

The FM modulator outlets have been combined according to the frequency plan, and have been routed towards the FM optical transmitters. The transmitters have been connected to patch panels located at optical cable bracing points. Finally, the patch panels have been connected to the cables.

At the hub, the operation is reversed. The fiber optic cables are connected to a patch panel which in turn is connected to FM fiber optic receivers. The optical receivers

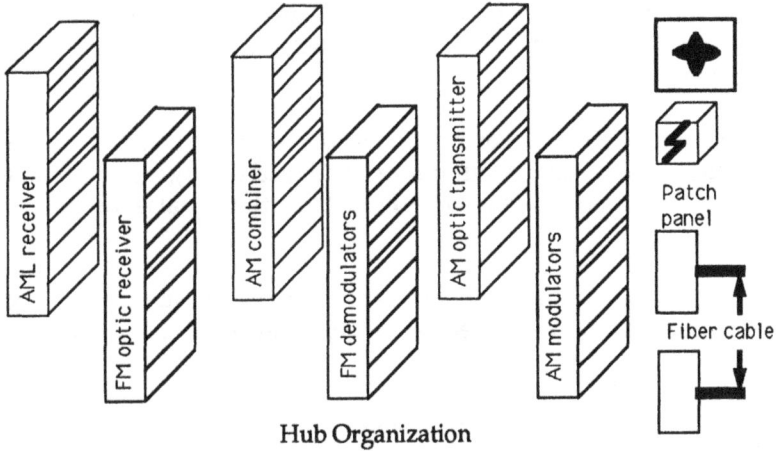

Hub Organization

output signals are routed toward the FM demodulation hardware, then to the AM modulation hardware. Finally, the signals are combined in groups of 12 and 18 channels, then routed toward the AM fiber optic transmitters. The transmitter outputs have been connected to a patch panel from which cables run towards subhubs.

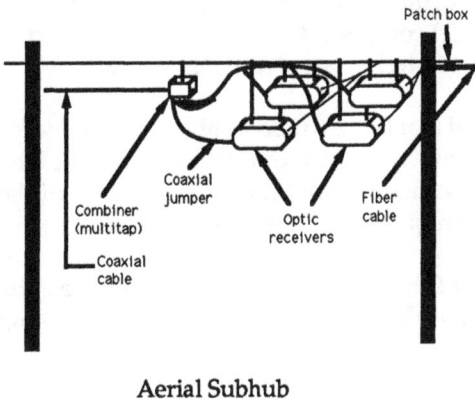

Aerial Subhub

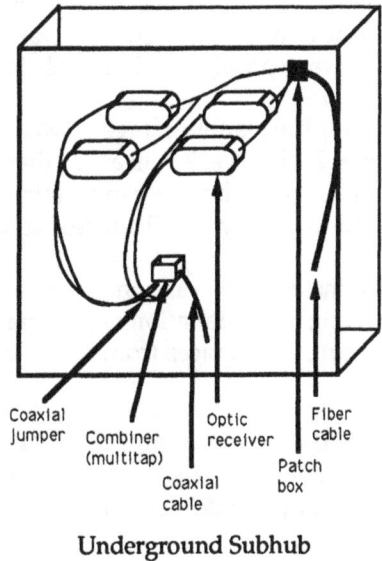

Underground Subhub

Since the technology used to interconnect the subhub is AM, there is no real equipment room or building involved. Hardware is located either on pedestals for underground installations, or they are attached to the strand, close to a pole in the case of an aerial installation.

The AM fiber optic receiver hardware is held in enclosures similar to trunk amplifiers housings. The receivers outputs are combined through a 4- or 8-way multitap according to the number of receivers. From there the signal is routed in the coaxial network.

The location of the subhubs has been carefully chosen in terms of accessibility, relation to the territory being served, safety regarding road accidents, theft, and vandalism.

Optical Fiber Installation

The link between the head-end and the hub is an underground link and uses our existing ducts. Since cables with 14 fibers are not regularly manufactured, a cable with 16 fibers was chosen.

The cable that was chosen has a steel strength member, with six loose fibers per tube, the latter being covered by a waterproof gel compound. The cable is surrounded by a waterproof tape. The cable is also protected by a polyethylene sheath.

During the planning stage, it was estimated that there would be about one splice for every 2.5 km, i.e. 7 splices for the 20 km link. However, due to the location of the manholes, the bending in certain conduit structures, and the cluttered state of the ducts along the run, 9 splices were in fact required to cover the 20 km run.

The installation was completed using all of the methods described above, e.g. by starting the installation not at one end but rather from nine different points along the run, at distances from 1 to 5 km.

From the hub to subhub no. 2, 7 km of fiber have to be buried. In fact, there are no pole lines or ducts which could help in reaching the given point without extending the run by an additional few kilometers.

The link requires a 10-fiber cable to route the signals. This cable is very similar to the one used to link the head-end with the hub. However, the cable only contains

10 fibers instead of 16. Also, as it will be exposed even more to rodents, a metallic tape was added to its outer sheath, just underneath the polyethylene coating. During cable installation, despite all the precautions taken to avoid cutting the cable, five splices were made rather than three.

From the main distribution center to subhub no. 7, 13 km of fiber must be installed on existing strands. When the link was being planned, it was decided that a cable with nine fibers would be installed. However, a 10-fiber cable was placed, since this is a regular count.

The cable that was chosen is a loose buffer tube, with a dielectric strength member. It was filled with a waterproof gel compound and its outer sheath is made of polyethylene.

In planning the run, it was decided that 6 splices would be required. However, only 4 were made, since it was possible to bypass most of the obstacles by pulling the cable using a nylon cord over a distance of about 1 km.

The total cost of the fiber, including installation and fusion splicing, is here different in every case. Installation hazards can also make a difference with the theoretical cost taken into account in the strategic study, when not well provisioned.

Operational Setup

Even before the splices were made and just before the cable splicing, the links were tested to gauge their performance levels. Some of the results appear in the table, "Fiber Optic Link Survey".

An analysis of that table shows that the installation was relatively well done. Losses in addition to theoretically estimated losses are only .25 dB, which is very good; in fact, .2 dB comes from the fiber itself. Only two splices generate losses greater than the basic manufacturer's losses.

As for reflections, most of the splices give very good results, though splice number 8 only gives a reflection level of -57dB compared to a basic level of -60 dB. This splice could therefore be redone, as it constitutes a weak point in the link.

FIBER OPTIC LINK SURVEY

Link Identification: _____ *Headend to Hub* _____

	Fiber Id.	Length	Attenuation	#	Splice Atten.	Refl.	Total Atten.	Refl.
Theoretical Calculation	1	2 km	.8 dB	1	.2 dB	-60 dB	1 dB	-60 dB
		2 km	.8 dB	2	.2 dB	-60 dB	1 dB	-60 dB
		2 km	.8 dB	3	.2 dB	-60 dB	1 dB	-60 dB
		2 km	.8 dB	4	.2 dB	-60 dB	1 dB	-60 dB
		2 km	.8 dB	5	.2 dB	-60 dB	1 dB	-60 dB
		2 km	.8 dB	6	.2 dB	-60 dB	1 dB	-60 dB
		2 km	.8 dB	7	.2 dB	-60 dB	1 dB	-60 dB
		2 km	.8 dB	8	.2 dB	-60 dB	1 dB	-60 dB
		2 km	.8 dB	9	.2 dB	-60 dB	1 dB	-60 dB
		2 km	.8 dB				.8 dB	
Total		**20 km**	**8 dB**		**1.8 dB**	**-50 dB**	**9.8 dB**	**-50 dB**
Installation		2 km	.8 dB	1	.2 dB	-60 dB	1 dB	-60 dB
		2 km	.8dB	2	.2 dB	-60 dB	1 dB	-60 dB
		.	.	.	.	.	.	.
		.	.	.	.	.	.	.
		.	.	.	.	.	.	.
Total								
Survey Date								
Total								
Survey Date								
Total								

Example of Link Survey Record

However, since the margins were estimated when the link was designed — a margin of 2 dB for the signal, for instance — while only .25 dB was actually used, the link may be considered operational. There remains a margin of 1.75 dB for possible repairs to the link and to cover wear and tear in the cable and splices.

As these first tests have been conclusive, the hardware may now be connected to the system.

The fibers are connected to the patch panels, and to the hardware. With an optical voltmeter, the output of all of the transmitters is adjusted to the rates in question, and with an optical reflectometer, surveys are taken for the receivers in order to test the system. The optical link is now operational.

Maintenance of the Link Network

Six months after the link's installation, performance tests are made on the link. This survey indicates some degradation of the system. Losses are 1 dB greater than what they were during installation, and reflections have increased by 2.8 dB. Therefore, it can be concluded that the link has a slight problem.

Instead of waiting for a system failure, the Hometown staff decided to check the fiber very carefully. Channels being routed on fiber number 1 are transferred to a fiber not being used outside of peak viewing hours. Even if the transfer takes only 2 minutes, because of the patch panels, it would be advisable for the transfer to take place at a time when the impact is as weak as possible.

Fiber number 1 is tested a few days later. Splice number 1 has a loss of .05 dB higher than what it had at installation time. This loss is not serious for the time being, but the splice in question will be checked carefully.

Splice number 3 also has a loss of .05 dB higher than at installation time, but the reflection rate has increased substantially. We would conclude that this splice must be remade before the fiber in question is put back into operation.

The 1.5 km length of fiber between splices number 3 and number 4 has a greater loss of .2 dB. Considering the length of the fiber, this loss is a substantial one. In order to check whether the entire cable has been damaged along this run, it has been decided that other fibers that were not used would be tested on this run. If this is the only fiber to show signs of being weak, it will be used again, but it will be carefully checked. If, on the other hand, other fibers show obvious signs of being weak, a portion of the link should be replaced. Tests performed by an optical reflectometer should provide answers as to where the weaknesses are located.

The same type of analysis is made on the length of fiber between splices number 5 and number 6, and the same measures are taken.

Splice number 8 also showed substantial damage. However, as it does not generate any more reflections as it did 6 months ago, it will not be replaced.

The additional losses of splice number 9 are not considered to be major, but this splice will also be carefully checked. Any subsequent damage will result in its being immediately replaced.

Conclusion

The Hometown Cablesystem optical network will most likely always meet the expected requirements, as its preventive maintenance is done on a regular basis. Also, the rate of failure should be very low, as action is taken immediately when something goes wrong.

References

1. Quotation from Northern Telecom Canada, Ltd.

2. Balmes, M., The Technology behind Heathrow, *Telesis*, Vol. 16, no. 2, 1989.

3. Average of several quotations.

4. Peach, D., *Multitier Specification for National Security/Emergency Preparedness (NSEP) Enhancement of Fiber Optic Long-Distance Telecommunication Networks*, National Telecommunications and Information Administration Report 87-226, National Communications System Technical Information Bulletin 87-25, US Department of Commerce, March 1988, Vol.2.

5. Fiberoptic Product News, *1988/89 Buying Guide*.

Chapter 6
Introducing Fiber Systems

6.1 The Strategic Plan

"Ultimately, vision gets translated into sales and profit growth and return on investment, but the numbers come after the vision" [1].

By now you certainly should have the "vision". Having the vision is believing the theology, respecting the mythology, and mixing in some adventurism, imagination, and a good dose of strategic planning. Fiber optics presents a challenge to the cable television industry.

Strategic Planning

You should be now in a good position to evaluate your new technology options and create a strategic plan to move your system into fiber optics. The strategic plan should pull together all the pieces of the puzzle and encompass current activities, near term (1-2 year) forecasts and longer term (3-5 year) objectives. The plan is built up year by year, using this book's methodology and optimized by year.

The plan should be a dynamic one. Fiber Optic technology is changing at a rapid pace with vendors coming out with new approaches seemingly every week. It is important to set up a process to create the plan in the first place, and also to review it periodically (probably every six months). The plan will have to evolve along with all of the things that influenced the plan in the first place; your system, your objectives, your competitors and the available technologies.

A strategic plan provides three main benefits to your business:

• it provides an overall road map to where you want to be in the future;
• it helps insulate your business from the day to day variations;

• it can reduce your engineering costs by narrowing choices and focussing resources.

While we have not quantified these benefits, it seems reasonable to assume that if you can shave 10-15% off the cost to engineer a rebuild, new build, or upgrade, then the strategic plan is probably a good investment.

The strategic plan is built on the following foundations:

• objective-setting,
• technology and architecture planning,
• engineering economics.

Objective Setting

Objectives for the system should be clearly understood. At the outset with Hometown Cablesystem we set some very specific objectives for channel capacity, signal quality, etc. These objectives have a very big impact on our near term plans. Often objectives may really be "subjectives"; that is, they are things we would like to do but are not quite sure how to quantify them. (For inspiration on this subject, see [2].)

Furthermore, many objectives relate to a variety of areas of system operation. In this book we have concentrated on the objectives that can be addressed directly by fiber optics. It is worth considering though that making a major change such as going to fiber optics will also have a large impact on the nonengineering parts of the business. For example, fiber can certainly be positioned positively from a marketing viewpoint and fiber will also have a positive impact on customer complaints (i.e. there should be fewer of them due to the higher reliability of the system).

There does not necessarily have to be just one set of objectives. There is probably only one set of the subjectives, but there will be many ways to map those into business strategies.

Technology and Architecture Planning

The technique of iterating between architectures and available technologies is applied for each year of the plan (and possibly for each set of objectives to be studied).

In interviewing fiber optic system vendors, we generally found a set of specifications applicable today, and another applicable six months or a year from now. These previsions can be factored into a multiyear plan. Beyond one year it can get pretty sketchy, but the worst case is if the vendors just meet the six month forecast and do not change after that. Since we want our investment to outlast the plan horizon, what is available today, and what is available 6-12 months from now are pretty much what we will have installed after the 3-5 years of the plan. Each time we do the plan, the available technologies will be updated.

It should also not be forgotten that there are other technologies than fiber optics available (we realize that is hard to believe). It could very well be that depending on what your objectives are and what your system looks like, that AM microwave radio or coaxial cable may just fit the bill. Analyzing the costs and implications of fiber optics on the specifics of your system may convince you to not use fiber optics, (although the authors wish to express some doubt on this subject as you may have expected).

Engineering Economics

To complete the plan and make sure all the cards are on the table, it is important to perform an engineering economic analysis of alternative scenarios. This involves calculating the net present value (NPV) of the plan (and its variations) against competing approaches [3].

Each scenario will have its own "optimal" plan using the methodology of this book. The one with the highest net present value is probably the best financial strategy to pursue. ("Probably" meaning that the economic analysis, as with any analysis technique, is a decision-making tool, not the decision itself. The decision is up to the decision-maker himself or herself - i.e. the cable system operator).

6.2 Technology Transfer

There are two aspects of technology transfer:

* training,
* documentation.

Training

We have found that to date most of the industry's fiber optic training has come from the vendor community. While some of it is very good, there are not too many philanthropists out there and they are likely really trying to sell you something. In fact, you may find during your fiber optic training, that you are really getting "unsold" on fiber optics altogether [4].

There are a number of university courses on fiber optics but these tend to be somewhat theoretical and quite expensive. Recently the Society of Cable Television Engineers has started offering workshops on AM Fiber Optics which support their certification program [5].

Training does not have to be overwhelming. You do not need a whole roomful of fiber optic "techies" in order to get into the technology, but being somewhat sufficient will help keep your engineering and ongoing support costs down.

Documentation

Again this has mostly come from the vendor community. As the penetration of fiber systems in CATV increases it will be necessary for the system operator to ensure he has adequate operating documentation.

Technology transfer is not something we have chosen to cover extensively in this book although there is considerable information of this nature in Chapters 3,4, and 5. For the planner, it is something to bear in mind when analyzing the options.

6.3 Field Trials

It is prudent particularly for the first application of a new technology to put the product into a "field trial" mode. This involves placing the equipment into live service while being heavily scrutinized. Field trials can last a few weeks or a few months depending on the scope of the trial. It should be planned to be sufficient for you to check all the technical and operational parameters of the system, as well, if possible, to have some subscriber reaction.

Field trials are best performed on production equipment as they will be in the field and "live". Initial bugs and early life failures should have already been sorted out by the manufacturer. The time and cost of a field trial should be included in your strategic plan.

6.4 Operations Training

A nuance on the training aspect of technology transfer is the training of the operations staff; i.e. the nontechnical personnel. These people from customer service representatives to accountants should understand the magnitude of the company investment and what it should mean to them, and to your subscribers. Because the CATV service is so tied to the coaxial cable plant there are undoubtedly many things that will change for the operations staff when you start going "fiber".

Making a significant investment in your plant can be a marketing tool and a community relations improvement vehicle. Do not neglect all the people who can have an effect.

6.5 Project Management

Once you have given the go-ahead to the strategic plan, it is necessary to manage the introduction of the new technology as you would any engineered job.

Managing the fiber project involves:

- Feasibility Study

You have a good start on this from the analysis done for the strategic plan, but specific feasibility studies may be needed to look at particular routes, types of plant, powering, civil works and other considerations.

- System Engineering

Your strategic plan took a first cut at the system engineering (link budgets, fiber count, etc.). Now you can do the detailed engineering portion of the job which includes the specific vendor calculations for the routes required. Plant records and material lists are part of this step.

- Project Coordination

This involves purchasing the goods, coordinating sites and deliveries, and managing the day to day operation of installing a system.

- System Installation

There are many vendors of installation services who can be contracted to marshall, install, splice, and acceptance test your fiber optic system. Larger CATV companies may choose to do this themselves. If so, the upfront costs of test gear, and splicing equipment should be factored into the cost equations.

- Ongoing Maintenance

Fortunately with fiber systems there is less ongoing maintenance than with coaxial systems. Since the input and output of the fiber system is the TV signal you are used to, your test equipment and training time is all preserved.

6.6 Hometown Cablesystem - The End of an Iteration

In the previous chapters we took a first stab at an architecture and some new technology for Hometown Cablesystem. We ended up with an FM trunk from the head-end to the hub location, and six AM fiber links to secondary hubs serving about 9000 subscribers each. Our choices are not yet optimized for Hometown as we have only completed one interation of the strategic plan. We did one round of technology selection to get the FM trunk, and one round of architecture analysis to get the six secondary hubs. We also found we could implement an alternate route infrastructure for the supertrunk for about $15 per subscriber.

To summarize, after the first iteration, we have:

Reference cost for the channel upgrade (i.e. with microwave and coax)	$70 per sub.
1st pass AM Fiber Trunk	-$ 3 per sub.
2nd pass FM Fiber Trunk	+$13 per sub.
Fiber Distribution (to 9,000 home level)	+$14 per sub.
Fibers Added for Futures	+$ 6 per sub.
Alternate route infrastructure for Trunk System	+$15 per sub.
Coaxial equipment not needed (i.e. included in the original $70 upgrade cost, but not needed now)	-$ 7 per sub
Total planned after 1st iteration	$108 per sub

At this point in the plan we are over the $100 objective for implementing the strategy. One area in question is the alternate trunk route. We have learned the following by looking at alternate route:

• The possibility of wanting an architecture with an alternate route reinforced the technology selection process. We had changed from AM to FM for the 20 km trunk. In fact, the alternate route in Hometown is about 45 km which is definitely beyond current AM capabilities.

• The alternate route is somewhat expensive. At $15 per subscriber, all we got was 45 km of cable and some rudimentary optical switches at each end to transfer the signals over. We still have not provided for a sharing or duplication of the lasers, which are in fact the more likely components to fail.

Our list of priorities for the second iteration would be:

• Technology Selection
>   -optimize hub for FM to AM transition,
>   -explore splitters for the secondary hubs.

• Architecture Analysis
>   -calculate reliability with and without the alternate route,
>   -look at protection or alternate route schemes for the secondary hubs,
>   -check our resultant system for possible new service applicability.

After the second iteration, we should have found an approach that would make us feel comfortable. We would then do a detailed comparison with the original budgets to see where there is room to make adjustments.

References

1. Naisbitt, J., *Reinventing the Corporation*, Megatrends, Ltd., 1985.

2. Naisbitt, J., op. cit.

3. There are a great number of texts and courses on Engineering Economics. Generally they involve using the concepts of interest and return on investment to evaluate the equivalent present cost of future capital and operating expenses. These expenses are offset by revenue to yield the "net present value". Different scenarios for deployment or new services can be evaluated and calculated using engineering economic techniques. The solution with the greatest net present value is the best financial strategy.

4. At one session we attended, we were told that the "mean time to repair" a fiber is longer than that of coaxial because of the complexity of fiber optic splicing. What they neglected to mention was that for the equivalent length of cable the mean time between failure is also much longer for the fiber.

5. For example, there was an AM Fiber Optics tutorial presented at the Cable-Tec Expo Show in Orlando in June 1989.

Index

The Artech House Telecommunications Library

Vinton G. Cerf, *Series Editor*

The Telecommincations Deregulation Sourcebook, Stuart N. Brotman, ed.

Digital Cellular Radio by George Calhoun

E-Mail by Stephen A. Caswell

The ITU in a Changing World by George A Codding, Jr. and Anthony M. Rutkowski

Design and Prospects for the ISDN by G. DICENET

Introduction to Satellite Communication by Bruce R. Elbert

A Bibliography of Telecommunications and Socio-Economic Development by Heather E. Hudson

Communication Satellites in the Geostationary Orbit by Donald M. Jansky and Michel C. Jeruchim

World Atlas of Satellites, Donald M. Jansky, ed.

Handbook of Satellite Telecommunications and Broadcasting, L.Ya. Kantor, ed.

World-Traded Services: The Challenge for the Eighties by Ramond J. Krommenacker

Telecommunications: An Interdisciplinary Text, Leonard Lewin,ed.

Telecommunications in the U.S.: Trends and Policies, Leonard Lewin, ed.

Introduction to Telecommunication Electronics by A.Michael Noll

Introduction to Telephones and Telephone Systems by A.Michael Noll

Integrated Services Digital Networks by Anthony M. Rutkowski

The Law and Regulation of International Space Communication by Harold M. White, Jr. and Rita Lauria White

Signal Theory and Processing by Frederic de Coulon

Digital Signal Processing by Murat Kunt

Principles of Secure Communication Systems by Don J. Torrieri

Digital Image Signal Processing by Frederic Wahl

Advances in Computer Systems Security: 3 volume set, Rein Turn, ed.

Techniques in Data Communications, by Ralph Glasgal

Measurement of Optical Fibers and Devices by G. Cancellieri and U. Ravaioli

Codes for Error Control and Synchronization by Djimitri Wiggert